EXTINCTION

EXTINCTION

A RADICAL HISTORY

EXPANDED EDITION

ASHLEY
DAWSON

OR Books
New York · London

Published by OR Books, New York and London
Visit our website at www.orbooks.com

All rights information: rights@orbooks.com

First printing 2022

Cataloging-in-Publication data is available from the Library of Congress. A catalog record for this book is available from the British Library.

ISBN 978-1-68219-299-3 paperback
ISBN 978-1-68219-294-8 e-book

Text design by Under|Over.

Cover art by Brandon Ballengée, *RIP Charles Island Tortoise: After Ernst Haeckel, 1904/2014.* Artist cut & burnt chromolithograph, etched glass funerary urn, & ashes (18 x 14 in). Photo by Casey Dorobek, courtesy of Ronald Feldman Fine Arts, New York.

TABLE OF CONTENTS

1: PREFACE TO THE EXPANDED EDITION

In the summer of 2017, members of the volunteer-run Entomological Society Krefeld in Germany published a paper that shocked both scientists and the general public. The work of the citizen scientists in Krefeld documented a nearly 80 percent decline in the number of insects caught over the course of several decades in special traps located in local nature reserves.[1] The speed and scale of the insect population crash stunned even professional entomologists who had been studying the problem in depth. Members of the Entomological Society Krefeld worked with scientists at Radboud University in the Netherlands on a trend analysis of the data they had gathered over the years, controlling for things like the effects of nearby plants, weather, and forest cover on fluctuations of insect populations. The final study, which looked at sixty-three nature preserves across 17,000 sampling days, found consistent declines across every habitat studied. The authors wrote that this suggested "that it is not only the vulnerable species but the flying-insect community as a whole that has been decimated over the last few decades."[2] News of this dramatic die-off of insects in what should have been a relatively stress-free environment rocketed around the world, generating headlines about an "insect apocalypse."[3] The shocking news indicated that something absolutely fundamental had gone wrong in the relationship of people to the natural world in the world's wealthy nations. Responding to a subsequent study of insect populations in Britain that confirmed the stark data from Krefeld, Gary Mantle, chief executive of a UK-based wildlife trust, argued that "this unnoticed apocalypse should set alarm bells ringing. We have put at risk some of the fundamental building blocks of life."[4]

Similar studies of mass die-offs in other portions of the natural world were published roughly contemporaneously with the work in Krefeld. For example, scientists at the Royal Botanic Gardens in Kew, England released a report on the state of the world's plants and fungi which revealed that two-fifths of the world's plants are at risk of extinction.[5] Worse still, in 2018 a new report by the WWF revealed that 60 percent of mammals, birds, fish, and reptiles have been wiped out since 1970.[6] But it is not just that quite unfathomably large numbers of specific species are at risk of annihilation. What these studies showed is that even when species are not threatened with extinction, their numbers are often dropping precipitously. What we're losing, in other words, is not just the *diversity* part of biodiversity, but the *bio* part as well. Even viewed in terms of sheer quantity, life on this planet is being liquidated at unprecedented rates. From the citizen scientists of Krefeld to the foremost academic experts, a grim alarm was being raised: the annihilation of wildlife is an emergency that now threatens human civilization.

But who or what is responsible for this crisis? One of the things that catalyzed me to write my book on extinction was the impression circulating in much public discourse at this time that human beings in general were responsible for the biodiversity crisis. Reading dominant news coverage of the biodiversity crisis as well as many scientific reports, one got the impression that human beings are like a plague of locusts, descending on pristine natural environments, stripping them bare, and then moving on, leaving nothing but barren earth in our wake. It seemed that there is something intrinsic to the human condition, perhaps as a result of our long evolutionary history as a species, that drives us to exploit the natural world in a completely unhinged manner. Or, in a more moralistic vein, that human beings are innately greedy and destructive. This analysis needed to be challenged, it seemed to me, since it was likely to induce political paralysis. After all, if human beings are inherently and immutably bent on plundering nature, why bother to fight against ecocide?

But beyond these questions of hope and political agency, dominant articulations of the biodiversity crisis also seem to me to be extremely racist and remarkably ahistorical. It is not humanity in general that is responsible for the crash of biodiversity: many Indigenous people, forest-dwelling peoples, and peasants around the world have existed for millennia in a remarkably balanced, even symbiotic, relationship with the natural world. In addition, historical analysis—including many of the studies I mentioned above—shows that extinction rates only really ticked up significantly during the period of European colonial expansion after the fifteenth century, increased markedly again following the Industrial Revolution, and then exploded during the so-called Great Acceleration after 1945. In other words, the biodiversity crash is a product of the intertwined forces of colonialism and capitalism. As I put it in my book, "by wrenching specific elements out of the complex ecosystems in which they are intertwined and turning them into commodities, capitalism remorselessly breaks down the natural world into impoverished but exchangeable forms, simultaneously discarding all those elements that don't appear to have immediate exchange value." The brutal contradictions of a capitalist system based on ceaseless, feckless expansion on a finite natural resource base are apparent in the woods, the streams, the trees, and the oceans as the planet and its precious, multitudinous life forms endure the Sixth Extinction.

Our understanding of the causes of this wave of global ecocide shapes the ways in which we think about resisting it. If it is the polluting behavior of a handful of powerful corporations and the unsustainable consumption patterns of certain people in specific parts of the world that are to blame for the biodiversity crisis, rather than humanity in general, then we may hope to stem and even reverse ecocide by calling out, reining in, and shutting down such unsustainable practices. If it is capitalism rather than human nature that is to blame, we must struggle to save the planet by putting an end to an economic system that encourages unfettered rapine. But where to start? While I do believe that environmental activists must embrace an anti-capitalist politics, I certainly do not intend my book to imply that we must wait for the overthrow of capitalism to challenge ecocide. There are many things that can be done here and now to address the biodiversity crisis. The members of the Entomological Society Krefeld model one possible reaction through their work as citizen-scientists. For the truth is that our knowledge of the natural world is still remarkably partial, and that longstanding prejudices still shape what kinds of life gets counted and valued. After all, there are plenty of studies of endangered charismatic mega-fauna like polar bears, but remarkably little scientific knowledge about the humble but absolutely essential earthworm or mycorrhizal fungi. Simply documenting these worlds under our feet and in the fields around us, as the members of the Entomological Society Krefeld did, can be a radical step, an initial salvo in a campaign against the handful of transnational agro-chemical industries that are one of the main forces driving planetary ecocide.

In addition, activist campaigns against extinction and ecocide must think in intersectional terms, seeing, publicizing, and fighting against the powerful corporate and financial organizations driving habitat destruction around the planet. This means challenging the efforts of powerful corporations and states to erode protections against toxic chemicals, including the US's recent effort to dismantle the precautionary principles in the European Union's constitution—principles which protect wildlife and citizens of the EU from all sorts of harmful chemicals.[7] More broadly, activists must think beyond narrowly-defined conservation campaigns. For example, movements fighting ecocide must trace the links between the biodiversity crisis, the use of deadly pesticides, land grabs, the expansion of new forms of extractivism, and the unsustainable model that underlies the global system of industrial agriculture. As the ecologists Ivette Perfecto, John Vandermeer, and Angus Wright show in their book *Nature's Matrix,* since biodiversity thrives in the matrix of diverse, non-toxic agricultural ecosystems maintained by small-scale peasant farmers around the world, the preservation of biodiversity is tied to the strength of movements for food sovereignty and agro-ecology.[8] Movements against ecocide thus need to support organizations like *La Via Campesina*, the global peasant confederation, since such organizations are the primary protagonists in the fight to sustain nature's matrix.

In addition, activists against extinction need to be careful about embracing some superficially conservationist programs advanced by powerful groups today. Often such initiatives actually contribute to the deterioration of natural environments while also dramatically worsening social conditions for some of the world's most vulnerable people. Take REDD+, the forest conservation program established under the aegis of United Nations climate negotiations.[9] As the Xapuri Declaration of forest dwellers and Indigenous Peoples argues, REDD+ is a form of carbon colonialism which allows polluting industries in rich nations to continue to spew carbon into the atmosphere while simultaneously privatizing forests in Global South nations, all in the name of conservation and the mitigation of carbon emissions.[10] REDD+ and similar schemes contribute to what might be termed conservation colonialism since these programs often lead to the ejection of Indigenous people and forest dwellers from their traditional lands.

Keeping these histories of carbon and conservation colonialism in mind, we should also be wary of schemes such as that of the American biologist E. O. Wilson to set aside half of Earth for safeguarding biodiversity.[11] Given the gravity of the climate emergency, the lure of such ambitious utopian schemes is strong. But whose land, we should wonder, will be set aside for such sweeping conservation

measures? Since Indigenous people constitute less than 5
percent of the world's population but protect 80 percent of its
biodiversity, the Half-Earth proposal seems likely to include
their territories.[12] Would Indigenous people be allowed to
remain on their land and retain sovereignty over that land in
such a global conservation scheme? All of this is to suggest
that questions of colonialism and imperialism need to be
kept in the forefront of debates about conservation if we are
to reverse histories and ongoing forms of genocide as well as
ecocide.

The last few years have hammered home the immediacy
of the biodiversity crisis. As researchers such as Rob Wallace
and Mike Davis have shown, COVID-19 emerged from
global capitalism's relentless drive towards expansion,
a drive that leads to deforestation and other forms of
extraction that are not just destroying fragile natural
environments around the world but also exposing humanity
to an increasing number of new and deadly pathogens.[13]
The pandemic has made it abundantly clear that the
biodiversity crisis should be of grave concern even to those
who seldom think about endangered populations of flora
and fauna. Of course this point is seldom made by racists
like former president Donald Trump, who would rather
blame the pandemic on Asian people than on the economic
system that so handsomely rewards him. This means that

activists and movements must fight hard to make sure the public understands the root causes—as well as the genuine solutions—of the biodiversity crisis. Unfortunately, as grim as the last years have been, COVID-19 is a foretaste of worse catastrophes to come if the world continues on its present course towards biodiversity and climate crisis. Deadly viruses are spreading across the planet on an almost annual basis, carbon continues to accumulate in the atmosphere and in terrestrial and oceanic ecosystems in dangerous quantities, and vulnerable ecosystems like the Amazon rainforest are on the verge of collapse. The science is clear: we're barreling towards global suicide, and taking most of the planet's precious ecosystems down with us. But it is the rich and powerful who are most responsible for our collective crisis, and it is the world's most vulnerable who will suffer—indeed, in many cases, are already suffering—first and most gravely. An escape from this catastrophic trajectory towards extinction will only come if the global majority rises up against the avaricious few who are benefiting, for the time being, from this increasingly deadly state of affairs.

2: A GLOBAL ATTACK ON THE COMMONS

His face was hacked off. Left prostrate in the red dust, to be preyed on by vultures, his body remained intact except for the obscene hole where his magnificent six foot long tusks used to be. Satao was a so-called tusker, an African elephant with a rare genetic strain that produced tusks so long that they dangled to the ground, making him a prime attraction in Kenya's Tsavo East National Park.[14]

These beautiful tusks also made him particularly valuable to ivory poachers, who felled him with poison arrows, carved off his face to get at his tusks, and left his carcass for the flies. The grisly death of Satao, one of Africa's largest elephants, is part of a violent wave of poaching that is sweeping the continent today. In 2011, twenty-five thousand African elephants were slaughtered for their ivory.[15] An additional forty-five thousand have been killed since that time. If the present rate of slaughter continues, one of the two species of African elephants, the forest elephant, whose numbers have declined by 60 percent since 2002, is likely to be gone from Africa within a decade.

The image of Satao lying faceless in the dust is a haunting one. While the elephant as a species will probably not go extinct (since some individuals are likely to be kept alive in game reserves and zoos), the decimation of their numbers in the wild reminds us of a broader tide of extinction, the sixth mass extinction Earth has witnessed. Only tens of thousands of years ago, during the Pleistocene epoch, Earth was home to an immense variety of spectacular, large animals. From wooly mammoths to saber-toothed cats to lesser-known but equally exotic animals like giant ground sloths and car-sized glyptodonts, megafauna roamed the world freely. Today, almost all of these large animals are extinct: killed, most of the evidence suggests, by human beings.[16] As they spread across the planet, *Homo sapiens* decimated populations of megafauna everywhere they went. Humanity essentially ate its way down the food chain when wiping out biodiversity.[17] Africa, our ancestral home, is virtually alone in harboring some remnants of the Pleistocene biodiversity. In the grisly death of Satao and his fellow elephants, we are witnessing the final destruction of the world's remaining megafauna, the endgame of an epoch of epic defaunation or animal slaughter.[18]

But it is not just charismatic megafauna like elephants, rhinos, tigers, and pandas that are being pushed to the brink of extinction. Humanity lives amid, and is the cause of, a massive decimation of global biodiversity. From humble invertebrates like beetles and butterflies to various terrestrial vertebrate populations like bats and birds, species are going extinct in record numbers. For example, since 1500, 322 species of land-based vertebrates have disappeared, and the remaining populations show an average 25 percent decline in abundance around the world.[19] Invertebrate populations are similarly threatened. Researchers generally agree that the current extinction rate is nothing short of catastrophic, clocking in between one thousand and ten thousand times the rate before human beings began to exert a significant pressure on the environment.[20] The Earth is losing about a hundred species a day.[21] In addition to this tidal wave of extinction, which conservation biologists predict will eliminate up to 50 percent of currently existing animal and plant species,[22] the abundance of species in local areas is declining precipitously, threatening the functioning of entire ecosystems.[23] This mass extinction is thus an under-acknowledged form—and cause—of the contemporary environmental crisis.

Although this wave of mass extinction is global, the vast majority of species destruction is concentrated in a small number of geographical hotspots. This is because biodiversity is unevenly distributed. On land, tropical rainforests are the primary nursery of biodiversity. Although they cover only 6 percent of the Earth's surface, their terrestrial and aquatic habitats harbor more than half the known species on the planet.[24] As E.O. Wilson puts it, the tropics are the leading abattoir of extinction,

their great verdant expanses chopped up into quickly
dwindling fragments, their plant and animal species
struggling to adapt to habitat destruction, invasive species,
overharvesting, and, increasingly, anthropogenic climate
change.[25] From the great Amazon basin, to the rainforests
of West and Central Africa, to the jungles of Indonesia,
Malaysia, and other parts of Southeast Asia, human beings
are eliminating the homes of millions of species. In doing
so, we are not only condemning vast numbers of species
(the great majority of them still unidentified) to extinction,
but we are also imperiling our own tenure on this planet.

With the publication of accessible works of science
journalism such as Elizabeth Kolbert's *The Sixth Extinction*,
the word has begun to get out about the dire plight of the
planet's flora and fauna. Kolbert's book takes readers on a
terrifying tour, interviewing botanists who follow the tree
line as it vaults up the side of mountains in the Andes and
marine botanists who track the acidification of the oceans.
The current wave of extinction, she explains, follows five
previous mass extinction events that have devastated the
planet over the last half billion years. This wave is predicted
to be the worst catastrophe for life on Earth since the
asteroid impact that destroyed the dinosaurs. Reflecting
on this melancholy reality, humanities scholars have begun
to write about "cultures of extinction."[26] In response to

such increasing concern, the Obama administration set up an interagency task force on wildlife trafficking, and began to discuss the trade networks linking the slaughter of elephants and rhinos to guerrilla groups and crime syndicates such as the Janjaweed and al-Shabab, which are using the high profits from the illicit wildlife market to fund their operations.[27]

All too often, however, initiatives such as Obama's result in a "war on poachers" that ignores the underlying structural causes that are driving habitat destruction and overharvesting of animals.[28] The planet's biodiversity hotspots, after all, are located in what Christian Parenti calls the "tropics of chaos."[29] In the planet's tropical latitudes, Parenti identifies a *catastrophic convergence,* a supremely destructive alignment of three factors: 1) militarization and ethnic fragmentation related to the legacy of the Cold War in postcolonial nations; 2) state failure and civil discord linked to the structural adjustment policies imposed on the global South by institutions like the World Bank in the name of debt repayment since the 1980s; and 3) climate change-fueled environmental stresses such as desertification. Parenti writes at length on the impact of this catastrophic convergence on postcolonial people and states, but the picture he provides of the stresses affecting the global South is incomplete without a consideration of the relations between humanity and the natural world in its fullest sense. We cannot understand the catastrophic convergence, that is, without discussing the decimation of biodiversity currently unfolding in the global South. Nor, conversely, can we understand extinction without an analysis of the exploitation and violence to which postcolonial nations have been subjected.

Extinction is the product of a global attack on the commons: the great trove of air, water, plants, and collectively created cultural forms such as language that have been traditionally regarded as the inheritance of humanity as a whole. Nature, the wonderfully abundant and diverse wild life of the world, is essentially a free pool of goods and labor that capital can draw on. As critics such as Michael Hardt and Antonio Negri have argued, aggressive policies of trade liberalization in recent decades have been predicated on privatizing the commons—transforming ideas, information, species of plants and animals, and even DNA into private property.[30] Suddenly, things like seeds, once freely traded by peasant farmers the world over, have become scarce commodities, and are even being bred by agribusiness corporations to be sterile after one generation, a product farmers in the global South have aptly nicknamed "suicide seeds."[31] The destruction of global biodiversity needs to be framed, in other words, as a great, and perhaps ultimate, attack on the planet's common wealth. Indeed, extinction needs to be seen, along with climate change, as the leading edge of contemporary capitalism's contradictions.[32]

Capital must expand at an ever-increasing rate or go into crisis, generating declining asset values for the owners of stocks and property, as well as factory closures, mass unemployment, and political unrest.[33] As capitalism expands, however, it commodifies more and more of the planet, stripping the world of its diversity and fecundity—think about those suicide seeds. If capital's inherent tendency to create what Vandana Shiva calls "monocultures of the mind" once generated many local environmental crises, this insatiable maw is now consuming entire ecosystems, thereby threatening the planetary environment as a whole.[34] There are at present no effective institutions to deal with the "cancerous degradation" of the global environment that David Harvey argues is brought about by capital's need for continuous exponential growth.[35] And yet capital of course depends on continuous commodification of this environment to sustain its growth. The catastrophic rate of extinction today and the broader decline of biodiversity thus represent a direct threat to the reproduction of capital. Indeed, there is no clearer example of the tendency of capital accumulation to destroy its own conditions of reproduction than the sixth extinction. As the rate of speciation—the evolution of new species—drops further and further behind the rate of extinction, the specter of capital's depletion and even annihilation of the biological foundation on which it depends becomes increasingly apparent.

Extinction: A Radical History is intended as a primer on extinction for activists, scientists, and cultural studies scholars alike, as well as for members of the general public looking to understand one of the great but all too often overlooked events of our time. Extinction is both a material reality and a cultural discourse that shapes popular perceptions of the world, one that often legitimates an inegalitarian social order. In order to respond adequately to this planetary crisis, we need to transgress the boundaries that tend to keep science, environmentalism, and radical politics separate. Indeed, extinction cannot be understood in isolation from a critique of capitalism and imperialism. *Extinction: A Radical History* begins with a discussion of the notion of the Anthropocene, using this term not simply to ask fundamental questions about when the sixth wave of mass extinctions began, but also about whom exactly is responsible for extinction. The second section outlines the different facets of extinction that are products of capitalism, from early modern forms of defaunation such as fur hunting to the episodes of mass slaughter such as whaling that arose in tandem with the industrial revolution. This section also discusses forms of *collateral ecocide* such as coral bleaching and extinction related to invasive species, as well as forms of *ecological warfare* such as the use of Agent Orange in Vietnam and the polluting of the Niger Delta. The third section of *Extinction: A Radical History* looks at *disaster biocapitalism*: the variety of political, economic, and environmental responses by capital to the extinction crisis. This section highlights not just the glaring failure of efforts to address extinction within a capitalist framework, but also the increasing trend to open a new round of accumulation using synthetic biology to address the crisis. Finally, the section on radical conservation explores various anti-capitalist solutions to the extinction crisis, approaches grounded in social and environmental justice.

The specter of extinction haunts the popular imagination today. Contemporary culture is filled with depictions of zombies, plagues, and other spectacular representations of ecological catastrophe.[36] For those who inhabit the wealthy nations of the global North, such representations are portents of a terrifying world to come. But for the billions of people around the world whom Ranajit Guha and Juan Martinez-Alier call "ecosystem people," whose fate is intimately intertwined with the planet's flora and fauna, the question of extinction relates directly to their own present and future survival.[37] The butchering of an elephant such as Satao may enrich a few poachers, but it dramatically impoverishes the ecosystem he inhabited. We are only just beginning to understand the impact of the liquidation of large wildlife like elephants on the habitats they inhabit, but it is becoming clear that such holes punctured in the web of life have a dramatic cascading effect.[38] As millions of species are snuffed out, the biodiversity that supports the planetary ecosystem as we and our ancestors have known it is imperiled. This catastrophe cannot be stemmed—let alone reversed—within the present capitalist culture. We face a clear choice: radical political transformation or deepening mass extinction.

3: **AN ETIOLOGY OF THE PRESENT CATASTROPHE**

"Gilgamesh listened to the word of his companion, he took the axe in his hand, he drew the sword from his belt, and he struck Humbaba with a thrust of the sword to the neck, and Enkidu his comrade struck the second blow. At the third blow Humbaba fell. Then there followed a confusion for this was the guardian of the forest whom they had felled to the ground. For as far as two leagues the cedars shivered when Enkidu felled the watcher of the forest, he at whose voice Hermon and Lebanon used to tremble. Now the mountains were moved and all the hills, for the guardian of the forest was killed."

—*The Epic of Gilgamesh* (2500–1500 BCE)

When did the sixth extinction begin, and who is responsible for it? One way to tackle these questions is to consider the increasingly influential notion of the *Anthropocene*. The term, first put into broad use by the atmospheric chemist Paul J. Crutzen in 2000, refers to the transformative impact of humanity on the Earth's atmosphere, an impact so decisive as to mark a new geological epoch.[39] The idea of an Anthropocene Age in which humanity has fundamentally shaped the planet's environment, making nonsense of traditional ideas about a neat divide between human beings and nature, has crossed over from the relatively rarified world of chemists and geologists to influence humanities scholars such as Dipesh Chakrabarty, who proposes it as a new lens through which to view history.[40] Despite its increasing currency, there is considerable debate about the inaugural moment of the Anthropocene. Crutzen dates it to the late eighteenth century, when the industrial revolution kicked off large-scale emission of carbon dioxide into the atmosphere.[41] This dating has become widely accepted despite the fact that it refers to an effect rather than a cause, and thereby obscures key questions of violence and inequality in humanity's relation to nature.[42] By thinking through the periodization of extinction, these questions of power, agency, and the Anthropocene become more insistent.

If we are discussing humanity's role in obliterating the biodiversity we inherited when we evolved as a discrete species during the Pleistocene epoch, the inaugural moment of the Anthropocene must be pushed much further back in time than 1800. Such a move makes sense since the planet's flora and fauna undeniably exercise a world-shaping influence when their impact is considered collectively and across a significant time span. Biologists have recently adopted such a longer view by coining the phrase "defaunation in the Anthropocene."[43] How far back, they ask, can we date the large-scale impact of *Homo sapiens* on the planet? According to Franz Broswimmer, the pivotal moment was the human development of language, and with it a capacity for conscious intentionality.[44] Beginning roughly 60,000 years ago, Broswimmer argues, the origin of language and intentionality sparked a prodigious capacity for innovation that facilitated adaptive changes in human social organization.[45] This watershed is marked in the archeological record by a vast expansion of artifacts such as flints and arrowheads. With this "great leap forward," *Homo sapiens* essentially shifted from biological evolution through natural selection to cultural evolution. Yet, tragically, our emancipation as a species from what might be seen as the thrall of nature also made us a force for planetary environmental destruction.

With this metamorphosis in human culture, our relationship to nature in general and to animals in particular underwent a dramatic shift. During the late Pleistocene era (50,000–35,000 years ago), our ancestors became highly efficient killers. We developed all manner of weapons to hunt big game, from bows and arrows to spear throwers, harpoons, and pit traps. We also evolved sophisticated techniques of social organization linked to hunting, allowing us to encircle

One of earliest recorded forms of creative expression by *Homo sapiens*, this Pleistocene stag was painted on the wall of the cave at Lascaux in southern France.

whole herds of large animals and drive them off cliffs to their death. The Paleolithic cave paintings of the period in places such as Lascaux record the bountiful slaughter: mammoths, bison, giant elk and deer, rhinos, and lions.

Some of the first images created by *Homo sapiens*, these paintings suggest an intimate link between animals and our nascent drive to imagine and represent the world. Animals filled our dream life even as they perished at our hands.

In tandem with this great leap forward in social organization and killing capacity, humanity expanded across the planet. From our ancestral home in Africa, we radiated outward, colonizing all the world's major ecosystems within the span of 30,000 years. We spread first to Eurasia, then, around 50- to 60,000 years ago, to Australia and New Guinea, then to Siberia and North and South America around 13,000 years ago, and then, most recently, to the Pacific Ocean Islands only 4,000 years ago. At the same time, humans underwent a massive demographic boom, expanding from a few million people 50,000 years ago to around 150 million in 2000 BCE. The late Pleistocene wave of extinctions cannot be understood in separation from this spatial and demographic expansion of *Homo sapiens*. In most places around the planet, the megafauna extinctions occurred shortly after the arrival of prehistoric humans.[46] On finding fresh hunting grounds, our ancestors encountered animals with no evolutionary experience of human predators. Like the ultimate invasive species, we quickly obliterated species that didn't know how to stay out of our way. The susceptibility of creatures who were unfamiliar with humans is evident from what biologists call the *filtration principle*: the farther back in time the human wave of extinction hit, the lower the extinction rate today.[47] This filtration effect means that in our ancestral home, Sub-Saharan Africa, only 5% of species went extinct, while Europe lost 29%, North America 73%, and Australia an astonishing 94%. Given the fact that biologists are only just beginning to understand the cascading, ecosystem-wide impact of the destruction of megafauna, it is hard to gauge the full impact of the late Pleistocene wave of megadeath. Nonetheless, given its planetary scale, the mass extinctions of the period are certainly the first evidence of humanity's transformative impact on the entire world's animal species and ecosystems.

When all the big game was gone, our ancestors were forced to find alternatives to their millennia-old hunter-gatherer survival traditions. Combined with climatic and demographic changes, the megafauna extinctions catalyzed humanity's first food crisis.[48] Pushed by these crisis conditions, humanity underwent what may be seen

as its second great transition: the Neolithic Revolution.
Given conducive environmental conditions—including
plant species that could be domesticated, abundant water,
and fertile soil—human beings shifted from nomadic to
sedentary modes of food production. This shift happened
remarkably rapidly, from about 10,000–8,000 BCE. The
transition to agriculture, with its greater capacity for food
production, led to a demographic explosion. About 10,000
years ago, around the time of the Neolithic Revolution,
the global human population was four million. By 5,000
BCE, it had grown to five million. Then, in a pivotal period
as settled societies developed on a major scale after 5,000
BCE, our population numbers began doubling every
millennium, to 50 million by 1000 BCE and 100 million 500
years later.[49] This demographic boom was accompanied by
the growth of settled societies, the emergence of cities and
craft specialization and the rise of powerful religious and
political elites. Paleontologists dub this period the Holocene
epoch, and it inaugurated an even more sweeping human
transformation of the planet than the previous wave of
extinctions. Indeed, the Neolithic Revolution must be seen
as one of the most fundamental metamorphoses not just
in human but also in planetary history. The domestication
of plant species and the exploitation of domesticated
animal power permitted human beings to transform
large swaths of the natural world into human-directed

agro-ecosystems. As "civilization" emerged, first in the city-states of Mesopotamia and then in Egypt, India, China, and Mesoamerica, humanity became a truly world-shaping species. Some critics have in fact dated the onset of the Anthropocene epoch from precisely this moment.[50]

The Neolithic Revolution also generated a fateful metamorphosis in humanity's social organization. Intensive agriculture produced a food surplus, which in turn permitted social differentiation and hierarchy, as elite orders of priests, warriors, and rulers emerged as arbiters of the distribution of that surplus. Much of subsequent human history may be seen as a struggle over the acquisition and distribution of such surplus.[51] Significantly, writing as a technology emerged in Mesopotamia during the fourth millennium BCE out of the need to record annual food production and surpluses.[52] The capacity conferred by cuneiform and subsequent systems of writing to transmit information and promote social organization clearly played an important role in the economic expansion of ancient societies. Indeed, writing appears to have emerged in tandem with the transformation of Mesopotamian city-states like Sumer into powerful empires.[53] Ancient Sumer generated an explosion of inventions that would be foundational to subsequent civilizations, including the wheel, the preliminary elements of algebra and geometry, and a standardized system of weights and measures that facilitated trade in the ancient world.[54] The Sumerians also pioneered less felicitous institutions such as imperialism and slavery.

As the idea of private property emerged and human society became organized around control over the surplus, writing also became a tool to record the resulting social conflicts. Much early writing, what we would today term literature, in fact documents chronic warfare. In works like *The Iliad* (760 BCE), for instance, we find what may be seen as a record of the intensifying warfare that accompanied the growth of city-states and empires.[55] The increased importance of warfare led to the rise of military chiefs; initially elected by the populace, these leaders quickly transformed themselves into permanent hereditary rulers across the ancient world.[56] Military values and a veneration of potentates came to suffuse ancient culture, at significant cost to the majority of the populace. While *The Iliad* celebrates the martial virtues of Greek warriors, for example, it also offers an extended lament for the violence unleashed as humans turned their skills of organized violence away from megafauna and onto one another.

The violence generated by what geologists call the Holocene epoch was directed not just at other human beings but also at nature. Indeed, what is perhaps humanity's first work of literature, the *Epic of Gilgamesh* (1800 BCE), hinges on a mythic battle with natural forces. In the epic, the protagonist Gilgamesh, not content with having built the walls of his city-state, seeks immortality by fighting and beheading Humbaba, a giant spirit who protects the sacred cedar groves of Lebanon. Gilgamesh's victory over Humbaba is a pyrrhic one, for it causes the god of wind and storm to curse Gilgamesh. We know from written records of the period that Gilgamesh's defeat of the tree god reflects real ecological pressures on the Sumerian empire of the time. As the empire expanded, it exhausted its early sources of timber. Sumerian warriors were consequently forced to travel to the distant mountains to the north in order to harvest cedar and pine trees, which they then ferried down the rivers to Sumer.[57] These journeys were perilous since tribes who populated the mountains resisted the Sumerians' deforestation of their land.

Ultimately, these resource raids were insufficient to save the Sumerian empire. The secret to the Sumerians' power was the creation of elaborate systems of irrigation that allowed them to produce crops using water from the region's two great rivers, the Tigris and the Euphrates.[58] Over time, however, the Sumerians' dams and canals silted up. Even worse, as the river water carried into fields by irrigation canals evaporated under the hot sun, it left behind its mineral contents, leading to increasingly saline soils. The only way to cope with this problem was to leave the land fallow for long periods of time, but as population pressure increased, this conservation strategy became impossible. Short-term needs outweighed the maintenance of a sustainable agricultural system. The Sumerians were forced, archeological records document, to switch from cultivation of wheat to more salt-tolerant barley, but eventually even barley yields declined in the salt-soaked earth.

Extensive deforestation of the region also added to the Sumerians' problems. The once-plentiful cedar forests of the region were used for commercial and naval shipbuilding, as well as for bronze and pottery manufacturing and building construction. As the *Epic of Gilgamesh* documents, the Mesopotamian city-states found themselves grappling with a scarcity of timber resources. The sweeping deforestation of the region also contributed to the secondary effects of soil erosion and siltation that plagued irrigation canals, as well as having a significant impact on the biodiversity of the region. As the Sumerian city-states grew, they were forced to engage in more intensive agricultural production to support the booming population and the increasing consumption of the civilization, with its mass armies and state bureaucracy. The Sumerians sought to cope with this ecological crisis by bringing new land into cultivation and building new cities. Inevitably, however, they hit the limits of agricultural expansion. Accumulating salts drove crop yields down more than 40% by the middle of the second millennium BCE. Food supplies for the growing population grew inexorably scarcer. Within a few short centuries, these contradictions destroyed ancient Sumerian civilization. The deserts that stretch across much of contemporary Iraq are a monument to this ecological folly.

Not all ancient societies went the way of Sumer. For about 7,000 years after the emergence of settled societies in the Nile Valley (around 5500 BCE), the Egyptians were able to exploit the annual flood of the Nile to support a succession of states, from the dynasties of the Pharaonic Era, through the Ptolemaic kings of the Hellenistic Period, to the Mamluk Sultanate, and the Ottoman Era. The stability of Egypt's agricultural system originated in the fact that the Nile Valley received natural fertilization and irrigation through annual floods, a process that the Egyptians exploited with only minimal human interference. Within decades of the introduction of dam-fed irrigation by the British in the nineteenth century, in order to grow crops like cotton for European markets, widespread salinization and waterlogging of land in the Nile Valley developed. The Aswan dam, begun by the British in the late nineteenth century, regulated the Nile's flood levels and thus protected cotton crops but undermined the real secret of Egypt's remarkable continuous civilization by retaining nutrient-rich silt behind the dam walls. As a result, the natural fertility of the Nile Valley was destroyed, replaced by extensive use of artificial, petroleum-derived fertilizers that placed Egypt even more deeply in thrall to the global capitalist economy.

This history of pre-modern ecocide is not intended to suggest that human beings are inherently driven to destroy the natural world upon which they ultimately depend. While it may be true that humanity's capacity to transform the planet on a significant scale through mass extinction dates back many millennia rather than simply two centuries, and that the Anthropocene therefore needs to be backdated substantially, it is only with the invention of hierarchical societies such as the Sumerian Empire that practices of defaunation and habitat destruction became so sweeping as to degrade large ecosystems to the point of collapse. The history of Egypt suggests that under the right material and cultural circumstances, human beings can achieve relatively sustainable relations with the natural world. It is the combination of militarism, debauched and feckless elites, and imperial expansionism, through which the Sumerians laid waste to much of the Fertile Crescent in pre-modern times, that renders ecocide so toxic as to destroy the very civilizations that carry it out. The collapse of ecocidal imperial cultures should serve as a potent warning to the globe-straddling world powers of today.

Ancient Rome offers additional stark evidence for the exploitative attitude towards nature that accompanies empire. One of the most striking characteristics of the early Roman Empire is its strong expansionary drive. Following a period of political conflict between patrician elites and plebeians (or commoners) in the 5th and 4th centuries BCE, large numbers of Romans began to migrate to newly conquered provinces. The treasuries of subjected lands such as Macedonia (167 BCE) and Syria (63 BCE) were looted, and a permanent of system of tributes and taxes was imposed, allowing taxes on Roman citizens to be eliminated.[59] This imperial expansion culminated in Augustus's conquest of the kingdom of Egypt, which allowed him to distribute unparalleled booty to the plebeians of Rome. He was the last emperor who could afford to do so.

In tandem with this looting of a significant portion of the ancient world, the Romans also used their conquests to deal with shortfalls in domestic agricultural productivity. First Egypt, then Sicily, and finally North Africa were turned into the granary of the empire in order to provide Rome's citizens with their free supply of daily bread. Deforestation caused by the Romans' agricultural enterprises spread from Morocco to the hills of Galilee to the Sierra Nevada of Spain.[60] Like the Sumerians, the Romans failed to engage in sustainable forms of agriculture, seeking instead to expand their way out of ecological crisis; the arid conditions that prevail across much of North Africa and Sicily today are testaments to their improvident and destructive approach to the natural world.

The people of Rome were kept obedient to imperial rule not just by subsidized grain, but by a combination of bread and circuses. In the latter, the class of slaves whose labor sustained the Empire was forced into gladiatorial matches to the death. They were joined in these bloody spectacles by wild animals brought from the farthest corners of the empire to die in combat with humans and with one another.

Lions, leopards, bears, elephants, rhinos, hippos, and other animals were transported great distances to be tortured and killed in public arenas like the Colosseum,

Roman mosaic from Veii (Isola Farnese, Italy) depicting an African elephant being loaded onto a ship, 3rd-4th century CE. Now in Badisches Landesmuseum Karlsruhe, Germany. Credit: Carole Raddato.

until no more such wildlife could be found even in the farthest reaches of the empire.[61] The scale of the slaughter was monumental. When Emperor Titus dedicated the Colosseum, for example, 9,000 animals were killed in a three-month series of gladiatorial games. While there is no evidence that the Romans drove any species to complete extinction, they did decimate or destroy numerous animal populations in the regions surrounding the Mediterranean Sea.[62] Indeed, the Roman Empire was probably responsible for the greatest annihilation of large animals since the Pleistocene megafauna mass extinction.[63] As was true of the Sumerians, Rome annihilated most of the large animals it could get its hands on and reduced most of the lands it conquered to desert.

To justify this carnage of wildlife, Roman attitudes towards the natural world shifted markedly. During the early days of the Republic, Romans regarded the Mediterranean landscape as the sacred space of nature deities such as Apollo, god of the sun, Ceres, goddess of agriculture, and Neptune, god of freshwater and the sea. As Rome expanded, however, these religious beliefs became largely hollow rituals, disconnected from natural processes.[64] During the high days of the empire, Stoic and Epicurean philosophies that legitimated the status-driven debauchery of the Roman upper classes prevailed. Orgies

Roman mosaic depicting abundant fish, fowl, fruits and vegetables
consumed at feasts.

of conspicuous consumption, in which the wealthy would
eat until they vomited, only to begin eating again, became
common. By the time Christianity became the official state
religion of Rome in the late 4th century, there was little to
differentiate Roman philosophy from the dominant attitude
of the Judeo-Christian scriptures, in whose creation myth
God grants human beings absolute dominion over the world
he has made. Humanity, the *Bible* and Christian tradition

held, was placed apart from nature by God, gifted with an immortal soul and a capacity for rational thought that legitimated the transformation of the natural world in the pursuit of human self-interest.

This orientation toward nature could not be sustained indefinitely. The spices and other luxury foods consumed by the dissolute Roman elite in their banquets had to be imported at great expense from locations as distant as India. The more exotic the food, the better; as recorded in the *Apicius*, a cookbook for elite Roman feasts, items such as thrushes and other songbirds, wild boars, raw oysters, and even flamingo were on the menu at elite banquets.[65]

Rome could not export enough goods to pay for these luxury imports, and was increasingly forced to pay with scarce gold and silver. Severe economic crises crippled the empire, forcing emperors after Augustus to end the customary distribution of free food to plebeians and to institute taxes on Roman citizens. The empire collected the funds it needed to subsidize military campaigns mainly from farmers, who consequently could not afford to invest in the production of crops and fell increasingly into debt.[66] Environmental degradation intensified, and the empire found itself unable to produce the food surplus on which its reproduction depended. Ultimately, Rome was no

longer able to pay its large and far-flung standing armies,
and, after a turbulent 500-year existence, the overextended
empire fell to the invading barbarian hordes of the north.
Rome today is remembered mainly for environmentally
destructive achievements such as the Colosseum, suggesting
that subsequent cultures learned remarkably little from the
unsustainable dominion and ultimate eclipse of the empire.

4: CAPITALISM AND EXTINCTION

Unpin that spangled breastplate which you wear,

That th'eyes of busy fools may be stopped there.

Unlace yourself, for that harmonious chime,

Tells me from you, that now it is bed time...

Licence my roving hands, and let them go,

Before, behind, between, above, below.

O my America! my new-found-land,

My kingdom, safeliest when with one man mann'd,

My Mine of precious stones, My Empirie,

How blest am I in this discovering thee!

To enter in these bonds, is to be free;

Then where my hand is set, my seal shall be.

—John Donne, "To His Mistress Going To Bed" (1654)

In the first of his accounts of his voyages to the New World, Christopher Columbus describes the island he named Española as an Edenic land whose "mountains and plains, and meadows, and fields, are so beautiful and rich for planting and sowing, and rearing cattle of all kinds, and for building towns and villages."[67] Greed and lust for power drip from Columbus's pen as he describes a marvelous land of abundant harbors and many rivers, "most of them bearing gold," and populated by naïvely generous inhabitants "so liberal of all they have that no one would believe it who had not seen it."

For Columbus, the biodiversity of this new world is equally notable, for, as he notes the islands are "covered with trees of a thousand kinds of such great height that they seemed to reach the skies," trees in which "the nightingale was singing as well as other birds of a thousand different kinds."[68] Columbus' breathless description of the material riches to be found in the "New World" set the tone for the European imperial expansion in the subsequent five centuries. As John Donne's sonnet to his mistress suggests, the lust for this imagined natural bounty was so strong that it permeated all aspects of European life, penetrating even the erotic fantasies of poets such as Donne. The flora and fauna of newly "discovered" lands appeared to Europeans to be

European representation of Columbus landing in the "New World," where
naive Indigenous people hand him their treasure in a sign of welcome.

an apparently boundless store of natural wealth, free for the
taking. Today we confront the baleful legacy of this feckless
appropriation and dissipation of the global environmental
commons.

If, in other words, human beings have engaged in notable
forms of ecocide throughout our history, it is only with
the expansion of Europe and the development of modern
capitalism that ecocide has taken on a truly global extent
and planet-consuming destructiveness. As Europeans
subjugated and colonized "virgin" lands, they dramatically

augmented processes of environmental degradation and extinction. The expansion of capitalist social relations through European colonialism and imperialism pushed what had previously been regional environmental catastrophes to a planetary scale. In addition, by transforming nature into a commodity that could be bought and sold, capitalist society shifted humanity's relations with nature into a mode of intense ecological exploitation unimaginable in previous epochs. Capitalism is not necessarily more immoral than previous social systems with regard to cruelty to humans and the gratuitous destruction of nature. As a mode of production and a social system, however, capitalism *requires* people to be destructive of the environment. Three destructive aspects of the capitalist system stand out when we view this system in relation to the extinction crisis: 1) capitalism tends to degrade the conditions of its own production; 2) it must expand ceaselessly in order to survive; 3) it generates a chaotic world system, which in turn intensifies the extinction crisis.[69] By wrenching specific elements out of the complex ecosystems in which they are intertwined and turning them into commodities, that is, capital remorselessly breaks down the complex natural world into impoverished but exchangeable forms, simultaneously discarding all those elements that don't appear to have immediate exchange value. In addition, as Marx argued in the *Grundrisse*, "capital is the endless and

limitless drive to go beyond its limiting barrier."[70] This
argument is quite clear on an intuitive level: any corporation
that doesn't outcompete its rivals will be driven out of
business in short order. Finally, as the era of globalization
demonstrates, capitalism creates a turbulent world in which
"all that is solid melts into air," as established modes of
governance and all other social forms are torn apart by a
gale of "creative destruction." While many commentators
have highlighted these dynamics of capitalism previously,
they are particularly starkly evident when seen through the
lens of extinction. These three key ecological contradictions
of capitalism are interwoven in practice, but their particular
dynamics are more evident when they are considered in
isolation, as they are in the following sections. The examples
discussed in these sections span the capitalist epoch, from
the earliest years of merchant capitalism to contemporary
forms of neoliberal globalization. Yet if these examples
suggest that the ecological contradictions of capital are
endemic, they also underline the remorselessly intensifying
character of capital's death-dealing reign.

CAPITALISM'S DEGRADATION OF THE ENVIRONMENT

The tendency for capitalism to degrade the conditions
of its own production is shockingly evident in the fur

trade, one of the main forces that drove European expansion after 1500. Aside from keeping wearers warm, fur clothes were status symbols in early modern Europe. The right to wear fur was tightly controlled by so-called sumptuary laws, which dictated that only people of certain social rank were allowed to don luxurious clothes. Nevertheless, as the mercantile bourgeoisie grew, the demand for furs spiraled. Western Europe quickly destroyed most of its fur-bearing mammals, and Russia began its long expansion eastward into Siberia, where it collected furs as tribute from conquered peoples such as the Tatars. By the sixteenth century, furs were the Russian state's largest source of income. Siberian beavers, sables, and martens were driven to the edge of extinction within two centuries.[71] The insatiable demand for fur consequently became one of the primary catalysts for European colonization of the Americas. Indeed, the French, Dutch, and English development companies established to facilitate European colonization of North America quickly realized that furs offered one of the most convenient means for the colonists to remit value back to Europe. Furs made fortunes for many European traders, who exchanged common and relatively cheap manufactured items such as iron axe heads with Native Americans for valuable beaver, deer, ermine, and other pelts.

European traders barter with Native American hunters for furs.

Over time, the Native American tribes caught up in the fur trade gradually abandoned their subsistent ways of life, becoming integrated into the emerging capitalist world system as specialized laborers working to harvest furs for European traders.[72]

In addition to transforming Indigenous subsistence culture, the fur trade catalyzed bloody conflicts between Native American tribes, including the so-called Beaver Wars of the mid-17th century, in which the Dutch- and English-backed Iroquois Confederation battled the predominantly Algonquin-speaking tribes of the Great Lakes region, whom the French supported. As beaver populations declined in places such as the Hudson Valley due to over-hunting, tribes like the Mohawks clashed with their neighbors to the north and west, where fur-bearing animals had not yet been hunted to the brink of extinction. The full human impact of these wars is still largely unknown since they took place beyond the frontier of European colonization, but they undoubtedly weakened the Native American tribes of the Northeast, making them more vulnerable to subsequent settler colonial campaigns of expropriation and extermination. In addition, such inter-imperial competition between the French and English led to higher prices for pelts, which increased the incentive for unsustainable over-harvesting of furs by European trappers and Native Americans. The fur trade continued until after the American Revolution, helping to make John Jacob Astor, owner of the monopolistic American Fur Company, into the US's first multi-millionaire. But Astor, having played a prominent role in decimating the continent's fur-bearing animals, abandoned the trade for speculation in real estate early in the nineteenth century. Although the beaver did not become extinct, its numbers were so reduced that it was no longer viable to hunt commercially. Scarcely two hundred years had passed since King Henry IV of France had granted the first charter to a European fur trading company in North America.

As Europe subjugated other parts of the planet, it dramatically transformed, and in most cases radically diminished, biodiversity of all kinds. In some cases, this was unconscious. The expansion of Europe into the Americas took the form of a great wave of novel biota, from smallpox and influenza viruses to pigs and horses.[73] Traveling alongside the European conquerors, these invasive species often wreaked havoc in the New World, killing many millions of Native Americans who had not been exposed to the new germs and transforming the landscape wholesale. In many cases, however, the Europeans also consciously obliterated biodiversity for their own selfish economic ends. For example, consider the plantation system. The immense diversity of the tropical and semi-tropical lands settled by the Portuguese and Spanish, early implementers of the plantation economy, was dramatically remade as land was turned over to grow a single crop such as sugar. As territories were subjugated and incorporated into European empires and the nascent capitalist system, indigenous agricultural practices that were adapted to the local climate (and consequently highly diverse and resilient) were extirpated. Such well-adapted agricultural practices were replaced by cash crops grown for export to the imperial metropole. Indigenous Peoples were displaced and slaves were imported to work the land, generating a brutal system of hitherto unequalled exploitation based on invented notions of racial difference.

In addition to displacing and killing many millions of people, the monocultures of the plantation economy quickly exhausted the land in the colonies, destroying soil fertility, and increasing vulnerability to pests.

By the late eighteenth century, plantation owners in the Caribbean islands had begun to worry about environmental degradation and climate change, which at the time was known as desiccation.[74] As a result of the deforestation linked to plantations, rain had ceased to fall on some of these islands. Mounting concern over the deteriorating environment led to the passage of the first conservation legislation, which set aside forest land in a forerunner of national park systems.[75] As plantation owners depleted the land, inter-imperial rivalry surged, with European colonial powers vying to capture islands whose fertility had not yet been depleted. The British abolition of slavery in 1833 can, in fact, be seen as a reaction to the declining productivity of its Caribbean plantations, rather than as an act of selfless humanitarianism.[76] Despite mounting awareness of the destructive social and environmental impact of the plantation system, however, the European powers continued to establish plantations around the world, as extensive tea, rice, and rubber industries were opened in Asia and Africa well into the twentieth century. The

Green Revolution of the second half of the twentieth century continued this trend towards displacement of small peasant agriculture by large landholdings devoted to export agriculture, with fossil fuel based fertilizers and pesticides used to cope with the resulting environmental stresses and contradictions.[77]

As Europeans colonized other parts of the world, they took cultural beliefs with them that legitimated their conquests. These ideologies of domination, intended to justify European expropriation of Indigenous people and their land, also established an exploitative attitude towards flora and fauna in the colonies. The English philosopher John Locke, for example, argued that God intended the land to belong to those who were "industrious and rational." These attributes were manifested in Europeans' "improvement" of the land through their labor, development work that, he argued, removed the land from its original communal state and made it the property of the Europeans. As Locke remarked, "He that in obedience to this command of God, subdued, tilled and sowed any part of [the land], thereby annexed to it something that was his property, which another had no title to…"[78] In other words, since Indigenous people weren't using the land properly, it didn't really belong to them and they could be dispossessed with no problem.

Not coincidentally, Locke owned plantations in English colonies in Ireland and Virginia.[79]

While part of the "improvement" that Locke envisaged was to come through the form of privatization known as *enclosure*, such development was also to take place through the application of modern science. As it was conceptualized by Francis Bacon and his followers in the seventeenth century, the scientific method involved interventions in a natural world represented as a female body, a body that had to be "twisted on the rack" and "tortured by fire" before it would reveal its secrets.[80] In

Women being burned at the stake in Europe, part of the campaign to enclose the commons that helped inaugurate capitalism.

many ways, Bacon and his acolytes were simply expanding on the Judeo-Christian tradition; after all, it is Adam whom God allows to name not just the animals in Genesis, thereby establishing his dominion over the natural world, but also Eve. But Bacon's representation of the forceful subjugation of a feminized nature also reflects a process of subjugation unfolding at the time he wrote: the violent acts of enclosure through which women, accused of being witches and often burnt at the stake, were deprived of control of their reproductive power in the early modern period.[81]

This subjugation mirrored the equally savage measures through which the European peasantry was expelled from the land they once held communally, as well as the bottomless depravity of colonialism and racial slavery, processes of expropriation, as Marx put it, "written in the annals of mankind in letters of blood and fire." As Bacon's account of scientific inquisition suggests, the scientific method took this reign of terror as one of its core metaphors, generating a model of patriarchal mastery over a passive feminized nature that set the terms for subsequent notions of progress through domination of the natural world. Doctrines of the objectivity and disinterestedness of the scientific method helped to obscure the potentially ecocidal, patriarchal, and racist character of techno-science, until the social movements of the late twentieth century arose to challenge science's role in legitimating colonialism, in depriving women of control of their bodies, and in creating deadly chemicals such as DDT.[82]

CAPITALISM'S CEASELESS EXPANSION

Capitalism is dependent on the conditions of production that it relentlessly degrades. By fecklessly consuming the environment, capital is figuratively sawing off the tree branch it is sitting on. But it does so because it must: it is a system based on ceaseless accumulation. Capitalists must constantly reinvest their accumulated profits if they are to survive against competitors, driving capital to expand at a compound rate.[83] Every limit to capital's expansion

European whaling took the industrialized slaughter of animals to the far reaches of the globe.

appears as an obstacle that it strives to overcome and fold into a new round of accumulation. But we live on a planet that is self-evidently finite. Capital's logic is consequently that of a cancer cell, growing uncontrollably until it destroys the body that hosts it.

The whaling industry is perhaps the best instance of this all-consuming drive to expand accumulation. Whales have endured the most prolonged and vicious attack by humans of any single species of animal.[84] Prior to the rise of capitalism, whales were hunted in sustainable numbers by Indigenous communities such as the Inuit in the Arctic, and by coastal-dwelling peoples such as the Basques, who intercepted immense but timid Bowhead and right whales as they made their annual migratory trek through the Bay of Biscay.[85] The Inuit and Basques killed whales in relatively limited numbers. But, as the industrial revolution took off, whales provided valuable commodities, including oil used for illumination and for greasing machinery in the factories of the period.

As a consequence, the growing markets of early modern capitalism exhausted stocks of coastal whales, and by the late seventeenth century whalers had to take to the open ocean in search of prey.[86] Maritime powers of the time such as the Dutch articulated a doctrine of freedom of the seas for their whaling fleets, opening the rich fisheries of the North Atlantic to commercial whaling by the competing European powers of the day.[87]

No efforts were made by the Europeans and their North American competitors to conserve stocks of whales. Instead, whalers acted as if their quarry was inexhaustible. Competition led to increasingly sophisticated techniques of slaughter, from the faster sailing ships of the late eighteenth century that hunted right whales to near extinction in several decades, to the invention in the mid-nineteenth century of the explosive harpoon gun and huge steam-powered factory ships, which allowed whalers to hunt faster fin and sperm whales in devastating numbers.[88] Although it was clearly in the industry's interest to limit the accelerating predation, the competitive dynamic of industrial capitalism made such forms of conservation impossible. Instead, whalers came up with far-fetched arguments to justify their monumentally shortsighted plunder of the oceans. For instance, in a chapter of *Moby Dick* entitled "Does the Whale's Magnitude Diminish? Will He Perish?" Melville's protagonist Ishmael ponders the question of the whale's extinction. Although he admits that whales were once far more easy to find in the oceans, he concludes that this is because whales now travel in bigger but less numerous groups, and that they have moved to the Poles in order to escape the whaling industry. As Ishmael's torturous reasoning suggests, whale populations had to be represented as limitless in order to justify the unsustainable competition of the industry. By the early twentieth century, humans had emptied the world's oceans of so many whales that commercial whaling was no longer a viable major industry.[89]

The decimation of whales and the crash of the whaling industry also illustrate the folly of the economic doctrines that grew up to legitimate capitalism. Adam Smith's *Wealth of Nations* (1776) is the clearest formulation of these doctrines. Smith believed that self-interested competition in the free market would generate beneficial outcomes for all by keeping prices low and creating incentives for a variety of goods and services. As Smith put it, "by pursuing his own interest [the individual] frequently promotes that of the society more effectually than when he really intends to promote it."[90] Private vices were purportedly transmuted into public virtues through the operation of what Smith described as the "invisible hand" of the market. Like many of his contemporaries, Smith believed in the inevitability of progress, which he assumed involved the production of greater material wealth. Yet, Smith's invisible hand completely ignored the issue of depletion and even extinction of such natural "resources" as fur-bearing animals and whales. In fact, classical economics is blithely ignorant of the impact of turning the earth's resources into capital, focusing only on the secondary problem of the distribution of resources between different competing ends.[91] But the earth's resources are not just scarce. They are finite. Like the whaling industry, classical economics is constitutively blind to this finitude, and consequently encourages both producers and consumers to use up resources as fast as possible in pursuit of greater profits and growth. Mainstream economics as formulated by Adam Smith and as practiced today celebrates values—selfishness, gluttony, competitiveness, and shortsightedness—that were once viewed as cardinal sins, and in the process provides intellectual justification for capitalism's disastrous pillage of the planet.

CAPITALISM'S CHAOTIC WORLD

If capitalism is based on the illusory hope that a mysterious "invisible hand" will reconcile ruthlessly self-interested competition with the common good, modern capitalist society is correspondingly organized around antagonistic nation-states whose competing interests, it is vainly hoped, will be attuned through various international forums. Yet, wracked by the periodic crises of over-accumulation that are a structural feature of capitalism, the bourgeoisie is impelled to seek markets abroad. Since their peers in other nations are driven to cope with system-wide crises through similar expansionary policies, the result is increasing inter-imperial competition and endemic warfare.[92] Capitalism thus generates a chaotic world system that compounds ecological crises.

In some cases, ecocide is a conscious strategy of imperialism, generating what might be termed ecological

European settlers proudly display skulls of slaughtered bison, carnage that was a key element in the campaign against Native Americans.

warfare. For example, the destruction of the huge herds of bison that roamed the Great Plains of North America was a calculated military strategy designed to deprive Native Americans of the environmental resources on which they depended.[93] When Europeans first arrived, the plains were inhabited by tens of millions of bison, providing Indigenous Peoples with resources that allowed them to maintain their autonomous, nomadic lifestyle. Commercial hunting of bison began in the 1830s, soon reaching a toll of two million animals a year.[94]

By 1891, there were less than 1,000 bison left on the continent, and the Native Americans had been crushed—defeated militarily and forced onto a series of isolated, barren reservations. Many of these reservations were subsequently turned into "national sacrifice zones" during the Cold War, when nuclear weapons were exploded in sites such as Nevada in order to perfect the US's military arsenal.[95] Similar ecological violence was meted out by the US military to other parts of the planet. During the Vietnam War, for instance, nearly twenty million gallons of pesticides were sprayed on the tropical forests of Vietnam in an effort to destroy the ecological base of the revolutionary Vietnamese forces. This virulent campaign of ecological warfare eventually generated a revolt among US scientists, who balked at what they called the systematic ecocide being

carried out by the military in Vietnam.[96] Despite this history
of war resistance, the US military, with more than 700 bases
worldwide, remains the single most polluting organization
on the planet.[97]

In many cases, however, animals and plants simply
suffer as collateral damage in the inter-imperial rivalries
generated by capitalism. In a system of competing capitalist
nations, no individual state has the power or responsibility
to counteract the system's tendencies toward ecological
degradation. Indeed, inter-imperial competition impels
individual states to shirk responsibility, seeking to score
points by blaming their competitors for failing to address
the environmental crisis. This fatal contradiction of
capitalist society has been abundantly evident in the rounds
of United Nations-sponsored climate negotiations during
the last two decades. During these negotiations, advanced
industrialized countries such as the United States and
Great Britain have refused to reduce their greenhouse gas
emissions significantly until developing nations such as
China, India, and Brazil offer to cut their emissions as well.
The industrializing nations respond by pointing out that
their per capita emissions are still far lower than those of the
wealthy nations of Europe and North America, and argue
that these countries have benefited from two hundred years
of industrial growth, effectively colonizing the atmosphere

to the exclusion of formerly colonized nations. As a result
of these antagonistic positions, no binding international
agreement on emissions reductions has been reached,
despite years of desperate pleas from scientists and civil
society. It is not simply that the climate and extinction crises
have arrived at a uniquely unpropitious moment when
neoliberal doctrines of financial deregulation, corporate
power, and emaciated governance are hegemonic.[98] Rather,
the deadlocked climate negotiations are a reflection of the
fundamentally irrational, chaotic, violence-ridden, and
ecocidal world system produced by capitalism.

Can capitalist society reform itself sufficiently to cope
with the extinction crisis? This is not simply unlikely.
It is impossible in the long run. While it is true that the
environmental movement did manage to push corporations
and the state into cleaning up local crises from the late 1960s
onwards, climate change and extinction suggest that the
capitalist system is destroying its ecological foundations
when viewed on a longer temporal scale. Recall that capital's
solution to periodic systemic crises is to initiate a new round
of accumulation. Capital essentially tries to grow itself out
of its problems. But, as we have seen, the extinction crisis
is precisely a product of unchecked, blinkered growth.
In such a context, conservation efforts can never be more
than a paltry bandage over a gaping wound. As laudable

as they are, conservation efforts largely fail to address the deep inequalities that capitalism generates, which push the poor to engage in deforestation and other forms of over-exploitation. Many of today's major conservation organizations were established in the last half of the twentieth century: the Nature Conservancy (1951), World Wildlife Fund (1961), Natural Resources Defense Council (1970), and Conservation International (1987). Yet during this same period, a new round of accumulation based on neoliberal principles of unrestrained hyper-capitalism has engulfed the planet. The neoliberal era has seen much of the global South become increasingly indebted, leading international agencies such as the World Bank to force debtor nations to harvest more trees, mine more minerals, drill for more oil, and generally deplete their natural resources at exponentially greater rates. The result has been a steeply intensifying deterioration in global ecosystems, including a massive increase in the rate of extinction.[99]

Despite this dramatic collapse of global ecosystems, the climate change crisis has unleashed a fresh round of accumulation, obscured by upbeat language about the investment opportunities opened up by the green economy. Neoliberal solutions to the climate crisis such as voluntary carbon offsets are not only failing to diminish carbon emissions, but are also dramatically augmenting the enclosure and destruction of the global environmental commons.[100] Such programs allow polluting industries in wealthy nations to continue emitting carbon, while turning the forests and agricultural land of Indigenous people and peasants in the global South into carbon dioxide "sinks" or biodiversity "banks." Under the green economy, vast numbers of people, plants, and animals are being sacrificed as collateral damage in the ecocidal exploitation of the planet. Capitalism, it is clear, cannot solve the environmental crises it is causing.

5: DECOLONIZING CONSERVATION

A specter is haunting the world—the specter of extinction. Faced with the unfolding, brutal destruction of biodiversity, the governments of wealthy nations, powerful transnational corporations, and big conservation organizations have proposed programs to cope with the crisis—but these efforts are organized around maintaining business as usual for the rich, while shifting the burden of environmental breakdown onto the backs of those whom the anti-colonial thinker and militant Frantz Fanon called the "wretched of the earth."[101] In the name of addressing the crisis, key players in the conservation industry like the WWF are adopting policies that amount to systematic forms of neo-colonialism. There is a seldom-acknowledge history behind these contemporary policies, a history grounded in racist colonial policies that systematically dispossessed and displaced Indigenous people from their lands across the globe.[102] Despite its carefully crafted image as the key means for saving the planet, conservation today is one of the leading edges of green capitalism, pioneering what I term *accumulation by extinction*.

As I argued earlier in this volume, capitalism is a system that is inherently destructive of the environment. It must expand continuously or be thrown into crisis. As an inherent part of doing business, capitalist firms swallow vast

quantities of natural resources and spit them out as waste—
detritus that includes the greenhouse gases that are driving
global warming.[103] But hasn't capitalism been around for
a long time? If it's so unsustainable, how has it managed
to stave off collapse? The geographer David Harvey has
pointed out that capitalism is periodically rocked by crises
provoked by the overaccumulation of capital, as businesses
and individual tycoons compete with one another to amass
the most surplus. The capitalist system tries to manage
these crises through what Harvey calls "spatio-temporal
fixes."[104] Put in simple terms, capitalists essentially have to
find a place to put the treasure they've hoarded through
successful competition. Failure to do so would mean their
capital might be seriously devalued, a wrenching experience
that economists term a recession or, worse still, a depression.
Capitalists evade such dangerous economic and political
crises by investing their money in new places, using what
Harvey terms processes of "spatial reorganization" like
urban development, as well as through violent forms of
geographical expansion such as imperialism. These kinds
of "fixes" also kick the can of crisis further down the road,
since the new sector where capital is invested eventually
grows into a bubble of its own which threatens to burst and
drag the economy in general down. The extreme inflation
of the housing market around the globe today is a prime
example of one of these spatio-temporal fixes: global elites

have plowed the mammoth assets they have stripped from ordinary people during the neoliberal era back into luxury condos in the sky, making housing unaffordable for ordinary people and threatening to provoke a crash as the worth of all these over-valued assets must be slashed dramatically.

Despite being faced with looming environmental breakdown, contemporary capitalists are doing everything they can to continue accumulating wealth. Indeed, capital accumulation is speeding up and intensifying in the teeth of the environmental crisis. According to the United Nations, for instance, the global material footprint rose from 73.2 billion metric tons in 2010 to 85.8 billion in 2017, an increase of nearly 18 percent.[105] While ramping up production and consumption, capitalists also seek to invest in sites around the world that are likely to absorb not just over-accumulated capital but also waste products of production such as greenhouse gases. This is what the conservation industry is increasingly specializing in: it sets aside so-called Protected Areas (PAs) to protect nature, and, not coincidentally, to absorb carbon emissions, allowing the capitalist system to carry on at full tilt.

Yet rather than leading to meaningful mitigation of emissions in particular and of the environmental crisis in general, such efforts to shift the crisis around in space and time are decimating the biosphere. Notwithstanding decades of global conservation work, annual high-level climate conferences, and proliferating rhetoric about greening capitalism, carbon continues to accumulate in the atmosphere and species still go extinct at ever-increasing rates. If conservation is a kind of spatio-temporal fix for the built-in contradictions that plague capitalism, it is one that actively produces extinction. For this reason, our present age can aptly be characterized as the era of accumulation by extinction.

Capitalists seldom openly admit that the system they benefit from works through accumulation by extinction. Instead, the dominant organizations of the capitalist world are doing their best to advance a series of false solutions to the planetary crisis. These proposals are designed to make it appear that elites and the system in general are responding thoughtfully and appropriately to the crisis, while nevertheless allowing them to carry on unsustainable business as usual. In most cases, these proposed green capitalist solutions actually intensify the problems they are supposed to rectify.

False solutions to the multifaceted environmental emergency are particularly apparent on the terrain of conservation. The proposals floated at the September 2021 meeting of the International Union for the Conservation of Nature (IUCN) in Marseilles are especially exemplary in their wrongheadedness. the IUCN is a kind of global conservation clearinghouse, established in 1948 and enjoying strong ties to other international organizations such as the United Nations and the World Bank, as well as to big conservation nonprofits like the WWF. As the IUCN put it in a press release following the conclusion of its 2021 World Conservation Congress in Marseilles, the IUCN is "setting the nature conservation agenda for the next decade and beyond."[106] The positions and policies adopted by the IUCN are the *de facto* dominant global responses to the extinction crisis. In addition, the IUCN operates beyond the terrain of species and habitat conservation, promising through its policies to address the climate crisis and to promote "sustainable development."

How exactly do elites propose to save an increasingly crisis-ridden natural world? In perhaps the most ambitious of the proposals to emerge from the Marseilles conference, the IUCN adopted the target of setting aside 30 percent of the planet as Protected Areas (PAs) by 2030. Motion 101, approved at Marseilles, calls on members of IUCN "to support, at minimum, a target of effectively and equitably protecting and conserving at least 30% of terrestrial areas."[107] The IUCN resolution responds to the recommendations of researchers, who in 2017 published a report as a companion to the Paris climate accord that called for a "global deal for nature," a key element of which was the plan to protect 30 percent of the Earth's surface and "sustainably manage" another 20 percent by 2030.[108] The 30 percent goal would constitute a doubling of the world's current Protected Areas. The IUCN is not alone in adopting the 30x30 goal: in October 2021, the Conference of Parties to the Convention on Biological Diversity was set to also adopt this goal.

It's important to note that the 30x30 goal is touted as a measure that will combat both the shocking loss of biodiversity recorded by researchers over the last half century *and* the climate crisis. A key measure advocated by the IUCN and by the conservation industry more broadly is the adoption of so-called nature-based solutions (NbS) to the climate and biodiversity crises. The 30x30 proposal, for example, promises to deal with both of these crises by banning human "interference" over vast areas of the planet, allowing nature there to regenerate and, in so doing, to absorb significantly increased amounts of carbon dioxide. It's the ultimate nature-based solution, something that its champions contrast to crackpot geoengineering schemes like shooting sulfur particles into the atmosphere or sucking carbon dioxide out of the air. What's not to like?

According to the activists who gathered at the "Our Land, Our Nature" Congress, a counter-summit to the IUCN's meeting in Marseilles, the 30x30 program and many of the other "nature-based solutions" advocated by the IUCN and the conservation industry more broadly are "colonial and false solutions."[109] Attendees at the "Our Land, Our Nature" Congress published a *People's Manifesto for the Future of Conservation* that offers a point-by-point debunking of dominant conservation strategies, as well as carefully thought-out alternatives to such hegemonic plans.[110] In what follows, I discuss *The People's Manifesto* in detail, explaining the context of its critique and the activism that informs its proposals for alternatives.

For the activists who drafted *The People's Manifesto*, the only "just, sustainable, and real solutions" to the biodiversity crisis begin with "Indigenous and local communities." After all, although Indigenous Peoples constitute only 5 percent of the world's population, they safeguard 80 percent of planetary biodiversity. The manifesto's insistence on "real solutions" underlines that there is an ideological and material battle playing out over how to cope with the loss of biodiversity and with the environmental crisis more broadly. On the one side of this conflict is the conservation industry, and on the other are Indigenous Peoples and local communities—all too often, the former perceives its interests as antithetical to those of the latter.

Embodied in powerful organizations such as the WWF, the conservation industry receives funding from well-meaning individuals in the wealthy capitalist nations, as well as from corporations and government organizations like the World Bank and USAID. The primary mode through which the conservation industry has worked historically is to cordon off land. For instance, the WWF's founding document, the *Morges Manifesto* (1961), makes clear the organization's purpose as a fundraising clearinghouse devoted to purchasing land: "They [existing conservation organizations] need above all money, to carry out mercy missions and to meet conservation emergencies by buying land where wildlife treasures are threatened."[111] There is a long colonial history behind such policies of acquiring land for "conservation." Environmental historians like Gregory Barton trace the roots of conservation to colonial policies in places such as India, where the "Charter of Indian Forestry" (1855) declared "annexation a ruling principle" under the British Raj.[112] Under the terms of the charter, any property not under private ownership could be designated as the property of the colonial state. Since the customary arrangements through which Indigenous Peoples, peasants, and other rural peoples across Asia and Africa commonly held and stewarded land tended to be invisible to European legal eyes, their land was conveniently available for annexation.[113] Similar laws were used to establish National Parks on Native American land in the United States.

Contemporary conservation organizations like the WWF enforce what at first glance seems to be a rigid separation between people and "nature." National parks and nature reserves are represented as spaces of a pristine nature, places saved from the forms of despoliation that are said to inevitably accompany human intrusion. This is where the idea of "fortress conservation" comes from: a Protected Area is declared by conservation groups; the people living there are ejected, often forcefully; fences and other forms of fortification designed to keep people out are put up, and "nature" is declared saved. But such exclusions are not universal: in reality, fortress conservation dispossesses colonized and racialized subjects while allowing wealthy white people virtually undisturbed access. These practices derive from colonial forms of dispossession. For example, small-scale hunting by tribal people was prohibited by the forestry charter in colonial India.[114] Colonized people who continued to hunt and gather on land annexed by the colonial state became "poachers," a term that is very much still in use to denote "illegal" hunting in nature reserves.

Significantly, the word "poaching" is derived from a Middle English word referring to the bag or pouch in which peasants would hide animals they'd killed on land where hunting was reserved for the Anglo-Norman aristocracy. A poor person caught poaching in medieval England was punished by death by hanging. Colonial-era organizations like the Society for the Preservation of the Fauna of the Empire (SPFE), founded in 1903 and often referred to as the first international conservation organization, maintained these traditions of exclusionary enclosure of the commons by establishing wildlife reserves where colonized people were not allowed to trespass but the British aristocracy could freely shoot trophy animals like tigers, lions, and elephants.[115]

It would be nice to think that these racist practices are a relic of the era of European colonialism. Lamentably, though, contemporary conservation remains informed by similar philosophical distinctions between people and "nature," and by nearly identical violent practices of ejection and interdiction of people in the Global South who are represented as "poachers." The philosophical foundation of conservation today leads it to embrace policies that are neocolonial and capitalist to the core. As the authors of the *People's Manifesto for the Future of Conservation* put it, "The mainstream approach to biodiversity conservation [. . .]

has largely been based on the flawed thinking that believes in 'nature' devoid of human presence. This single-minded focus has led to a model of conservation that is often violent, colonialist, and racist in approach—seizing and militarizing the land, criminalizing and destroying the ways of life of Indigenous and local communities, while destroying their knowledge." Today, many conservation organizations have embraced shoot-on-sight and even shoot-to-kill policies against "poachers," in a development that critics have described as the rise of militarized conservation.[116]

Not all contemporary conservation is militarized. Responding to the criticism of human rights groups, some conservation organizations have tried to establish community-based wildlife conservation programs. But in a discussion of a variety of these programs in countries like Namibia, Tanzania, and Zambia, a group of researchers concluded that while some of these programs do generate benefits for local communities, they disproportionately reward wealthier individuals and groups, and can even intensify inequality and exacerbate the marginalization of already-struggling groups.[117] For this reason, the authors of the *People's Manifesto* argue, "These attempted reforms are routinely manifested as window dressing because they focus on the implementation of a single project, while failing to acknowledge that the problem is structural. The underlying model of conservation is thus still far from being rights-based." In addition, the vast majority of funding for conservation continues to flow not into community-based conservation but into the kinds of fortress conservation that criminalizes local people. For instance, according to the journalist Mark Dowie, USAID spent nearly $300 million on Protected-Area-based conservation efforts during the 1990s; the five largest global conservation organizations, including the WWF, Conservation International, and The Nature Conservancy got more than 70 percent of that funding, while Indigenous Peoples' organizations got none.[118]

The global impact of fortress conservation is truly breathtaking. As Mark Dowie reports, there were only 1,000 official Protected Areas (PAs) in 1962, when many of today's big conservation organizations did not yet exist. Today there are 108,000 PAs around the world. According to Dowie, "the total area of land now under conservation protection worldwide has doubled since 1990, when the World Parks Commission set a goal of protecting 10 percent of the planet's surface."[119] This goal has been dramatically surpassed: according to the 2020 *Protected Planet* report, at least 17 percent of terrestrial and inland water, and 10 percent of coastal and marine areas, are conserved.[120] Since 2010 alone, Protected Areas covering almost 21 million square kilometers have been added. If this land were contiguous, it would be larger than the Russian

Federation. For conservation organizations, this is a great victory. But extinction rates continue to tick up around the planet. The answer, for big conservation, is to turn more land into Protected Areas. Hence the 30x30 goal endorsed at the Marseilles meeting of the UICN. This goal constitutes a near doubling of the already-dramatic amount of land in PA status.

Successfully meeting this goal for expansion of PA's to 30 percent of the Earth's surface would dispossess and displace roughly 300 million people, according to a recent report.[121] Reaching the goal of setting aside half the planet as Protected Areas, as renowned scientists such as E.O. Wilson have proposed, would directly affected over one billion people.[122] Such shocking estimates of the potential impact of conservation policies are important since, as Mark Dowie notes, hardly any national governments or international agencies keep records of the number of people displaced by the expansion of land dedicated to conservation. In an indicative case, however, Dowie mentions that when the African nation of Chad expanded the amount of land under protection from 0.1 to 9.1 percent in the 1990s, an estimated six hundred thousand people were turned into refugees.

The battle against such displacement is longstanding. Fiore Longo, research and advocacy officer for Survival International, told me that her organization has been campaigning to prevent evictions for three decades, beginning with a protest letter sent to the WWF in 1989 about people displaced by a PA in the African nation of Cameroon.[123] Longo and her colleagues noticed that such cases of dispossession—and attendant outcry from affected communities—were multiplying. As her organization conducted interviews with PA-affected communities, Longo and her colleagues began to realize that these forms of displacement were baked in to the conservation ethos, and consequently are systemic to the industry.

The big conservation organizations of the world have attracted serious economic support, not just from wealthy philanthropists moved by the plight of wildlife but also from large capitalist corporations. For the latter, supporting the creation of Protected Areas functions as a form of greenwashing, making it appear that they are addressing the climate crisis while diverting attention away from their continuing pollution of the environment. For example, the French company Total—the world's seventh-largest oil producer—announced plans to plant a "40,000 hectare forest" in the Republic of Congo in 2021.[124] The company says that the project will sequester more than 10 million tons of CO_2 over 20 years, making it a prime example of Total's commitment to "the development of natural carbon sinks in Africa."[125] The announcement follow's Total's creation of a Nature-Based Solutions unit in 2019, a development mimicked by other big oil companies like Shell, Eni, and Equinor.

Conveniently for these companies, such initiatives allow fossil capitalist corporations to continue polluting: Nature-Based Solutions like the "forest" in Congo are supposed to absorb carbon, thereby "offsetting" ongoing pollution of the atmosphere through the extraction, refining, and combustion of petroleum that is remains the core business of oil companies. Total's effort to greenwash its continuing destruction of the environment is particularly duplicitous since, as Simon Counsell reports, its NbS announcement came hard on the heels of the company's acquisition of an oil exploration permit in a 1.5-million-hectare patch of the Congo—a concession that lies beneath what was recently revealed to be one of the largest natural peat deposits in the tropics, a place that is among the most carbon-dense ecosystems on the planet. Not only would Total destroy this carbon-rich ecosystem in order to extract polluting petroleum, but the forest it would replace this precious ecosystem with is slated to be comprised of acacia trees, a fast-growing species native to Australia that will be harvested by timber processing facilities to be built in the area. So it turns out that the "forest" is actually an intensive wood farm, an enterprise to be run by a shadowy company with political connections to Congo's long-term dictator, Denis Sassou-Nguesso.[126]

The insistence that Indigenous and local people are the most appropriate protagonists of conservation in *The People's Manifesto for the Future of Conservation* is intended to overturn these traditions of greenwashing and exclusionary conservation, as well as to acknowledge the fact that it is Indigenous people and local communities rather than big conservation organizations who are the true stewards of global biodiversity. As political ecologist Prakash Kashwan and his colleagues note, "many of the world's poorest and most vulnerable (and nonwhite) populations host some of the world's most biodiverse areas, in contrast to, say, Western Europe, where many of the native forests and wild animals no longer exist."[127] Four out of the world's five High-Biodiversity Wilderness Areas, places where a minimum of 70 percent of their original vegetation remains intact—are in (formerly colonized) Global South nations.

The environmental historians Ramachandra Guha and Joan Martinez-Alier use the term *ecosystem people* to describe "communities which depend heavily on natural resources in their own locality."[128] Guha and Martinez-Alier contrast these communities with what they call *omnivores,* or individuals and communities with the "power to capture, transform, and use natural resources from a much wider catchment area, sometimes the whole world."[129] It stands to reason that ecosystem people would figure out ways to coexist and even nurture biodiversity since their continued existence essentially depends on maintaining the local resource base. Unlike omnivores, that is, they are not free to lay waste to a specific locale and then move on, or to stay comfortably in place and siphon resources from distant regions. Given the role of the roughly two billion people who live in and maintain the world's biodiversity hot spots, it makes sense to conceptualize these people and the multispecies networks they help sustain as *biodiversity communities*.

The *People's Manifesto* insists that the rights of these biodiversity communities be respected. It calls on the international community to agree to a complete halt of all PAs which exclude Indigenous and local communities, and to commit to respecting collective customary land and forest use by such communities. Consistent with the demands of Indigenous movements around the globe, the *Manifesto* argues that governments and conservation organizations must not embark on any conservation projects without the full Free, Prior, and Informed Consent of communities to be affected by such projects.

In addition, the *Manifesto* argues for reparations for the harms done to these communities through conservation policies of the past. This must include turning protected land back over to its historical and local guardians. The *Manifesto* does not lay out exactly how such forms of restitution and reparation would be carried out, but its insistence that a history of dispossession must be not just renounced but rectified through concrete policies is key to the project of decolonizing conservation. I make some suggestions about what such reparations might look like in the following chapter.

Addressing the extinction crisis successfully will also require upending the dominant ethos in conservation. As the authors of the *People's Manifesto* put it, "We need a conservation model that fights against the real causes of environmental destruction and is prepared to tackle those most responsible: overconsumption and exploitation of resources led by the Global North and its corporations." Conservation thus needs to incorporate a program of degrowth for the Global North, a paradigm and policy shift away from the constantly expanding production, consumption, and wastage of unnecessary products that characterizes advanced capitalist societies.[130] This does not imply that production in general should cease, since people's basic needs must still be satisfied and the infrastructures necessary for a transition away from fossil capitalist society built. But it does mean that the production of an endless deluge of manufactured needs and piles of unnecessary stuff must be brought to an end.

The omnivores of the world need to change how they live. If the world is to evade ecosystem collapse, wealthy people in the world's core capitalist nations but also the rich in Global South countries must reign in their consumption and learn how to subsist on resources drawn predominantly from the bioregions in which they reside. It is their excessive consumption rather than the behavior of biodiversity people that is responsible for the extinction crisis around the world. To make this shift will require challenging not just the social and economic power of elites but the entire capitalist system, with its reliance on global assembly lines and rampant exploitation of low-wage, feminized labor. Simply setting aside a portion of the planet as Protected Areas—aside from the injustice of such measures—assumes that capitalism will happily limit itself to the other portions of the world. But we know that capitalism is a system that is inherently driven to expand or collapse.

The forms of desire through which people in capitalist societies are encouraged to cope with their gnawing sense of alienation and loneliness through the purchase of heaps of material things must be transformed into genuine social connection and engagement in the building of new worlds. This should not be represented as a form of privation but rather as an opportunity to rethink and rework the barren worlds created by capitalism, to embrace the forms of collective becoming and world-making that radicals of the past have termed "communal luxury."[131] We know where to look for such alternative ways of being in the world: to precisely the communities most threatened by the capitalist system and the conservation industry. If biodiversity communities composed of Indigenous people, forest-dwellers, and other locally based groups in the Global South are the key stewards of the natural world, then it follows that the values and outlook towards the natural world that ground their way of life should not just be respected but also emulated.

6: ANTI-EXTINCTION

A living organism, after all, was a ready-made, prefabricated production system that, like a computer, was governed by a program: its genome. Synthetic biology and synthetic genomics, the large-scale remaking of the genome, were attempts to capitalize on the facts that biological organisms are programmable manufacturing systems, and that by making small changes in their genetic software a bioengineer can effect big changes in their output.

—George Church and Ed Regis, *Regenesis*

September 2014 was the fortieth anniversary of the US Wilderness Act, a law which protected millions of acres of public land and also provided a lucid definition of wilderness: "an area where the earth and its community of life are untrammeled by man, where man himself is a visitor who does not remain." Coinciding with this anniversary, however, the World Wildlife Fund released its *Living Planet Report*, which contained the shocking news that the number of wild animals on Earth has halved over the past forty years.[132] Clearly the strategy of setting aside dwindling islands of wilderness as "untrammeled" reserves, an approach central to wildlife conservation, is failing miserably. The massive wave of defaunation that has swept the globe over the last half century challenges the very idea of an unspoiled nature. There is no safe refuge from anthropogenic extinction. Indeed, the wilderness that remains has been so significantly degraded that we suffer from what J.B. MacKinnon calls *change blindness:* as the planet's remaining wilderness is degraded, each generation grows up with an increasingly impoverished view of natural biodiversity, so that human experience itself is undergoing a form of extinction.[133] What remedies can we imagine to reverse this natural and cognitive destitution? Does the specter of the end of speciation promise to catalyze a renewed restoration ecology? And, if so, what kind of wilderness should be resurrected?

If human beings are the prime authors of extinction, we can, according to advocates of *rewilding*, also be the creators of fresh wilderness. Introduced in the late 1990s by biologists Michael Soulé and Reed Noss, rewilding acknowledges the crisis in conservation provoked by dramatic defaunation. The concept was based on the then-radical idea that large, wide-ranging, usually carnivorous animals play a key role in preserving the diversity and resilience of ecosystems. In most cases, these keystone species, once viewed by human beings as a direct threat, have been displaced or driven to the edge of extinction. Rewilding entails the restoration of huge tracts of wilderness through the creation of large, linked core Protected Areas and the reintroduction of keystone species into such new wilderness. As imagined by its advocates, rewilding would not replace traditional conservation measures intended to protect the existing indigenous species of particular bioregions, but would complement such efforts by seeking to restore levels of biodiversity that have been eradicated from such sites in recent centuries.

The idea of rewilding has gained significant traction as a result of the successful reintroduction of wolves into Yellowstone National Park. Seen by the European settlers who colonized the region in the nineteenth century as destructive predators whose behavior destroyed "more

Feral wolves reintroduced into Yellowstone National Park. Credit: Steve Jurvetson via Wikimedia Commons.

desirable" species like deer and elk, wolves had been almost entirely exterminated in the lower 48 states of the US by the mid-1900s. In the 1960s, however, the National Park Service moved away from an anthropocentric policy of treating Yellowstone like a carefully controlled game reserve. Henceforth, the park's wildlife was to be allowed to manage itself. In response to this shift, biologists argued that wolves needed to be returned to Yellowstone in order to return the park's ecosystem to its "natural" condition, or at least to conditions before the arrival of European settlers, their cattle, and their predator-extermination campaigns. The idea of unleashing packs of wolves in Yellowstone generated a public outcry, but the reintroduction program, begun in the mid-1990s, has been a significant success. Yellowstone's gray wolves prey primarily on elk but also increasingly on bison, leaving carcasses that provide food to many other animals, including grizzly bears and cougars, helping to increase the numbers of these species. The wolves have driven elk herds out of the park's lowlands, leading to significant reforestation. As a result, record numbers of birds have returned to the park. Fish populations have also increased, as decreased grazing by elk has increased vegetation on riverbanks. Wolves are thus responsible for *trophic cascades*—chains of beneficial effects set off when an ecosystem's top predators change not just the numbers of their direct prey but also species with which they have no direct link.[134] Reintroducing predators and large herbivores in sites such as Yellowstone generates changes that cascade down the links of the ecosystem, transforming even the soil composition and atmosphere of the region. By catalyzing a notable increase in the park's biodiversity, Yellowstone's wolves have given flesh to the hopes of rewilders.

For advocates like George Monbiot, rewilding promises to restore not just wilderness ecosystems but also humanity's hope about the environment. We no longer need think of ourselves as simply seeking to preserve an increasingly impoverished natural world, as traditional conservation biology does, Monbiot argues. Ecological change need not proceed as a remorseless downward spiral towards the end of speciation. As Monbiot puts it, by reversing destructive processes, rewilding holds out the hope that our silent spring may be succeeded by a raucous summer.[135] While it may be true that a species, once extinct, is gone forever, ecosystems themselves can be regenerated through the reintroduction of megafauna. Rewilding thus suggests that we can reverse the flow of ecological time. It proffers a possible restoration of lost environmental time. This temporal shift also augurs a rekindling of human wildness, as our ideas of what nature should be are transmuted through the reintroduction of displaced or extinct keystone species such as the wolf.

Rewilding also promises to rework environmental space. Looking forward to the calamitous impact of climate change, biologists have proposed radical new doctrines with the ominous-sounding names *assisted colonization* and *ecological replacement.* As habitats are transformed by climate change, the static spatial boundaries of existing parks and refuges are likely to strand animal and plant species in increasingly unsuitable sites. The pace of change in this regard is shocking: some plants are literally running up mountains at the rate of tens of feet

A Woolly Mammoth, the most charismatic of megafauna and fetish object
of schemes for de-extinction. Credit: Tracy O. via Wikimedia Commons.

per year in order to cope with climate change-induced
habitat change.[136] Under such conditions, fears about the
negative impact of invasive species must be tempered
by the need to sustain entire ecosystems threated with
annihilation.[137] Assisted colonization responds to this
mutation of habitats by relocating endangered species
to new, ecologically appropriate reserves.[138] Ecological
substitution, conversely, involves introducing appropriate
substitute species to restore an ecological role that has
been lost when an original indigenous species goes
extinct. In responding to the increasing instability of
habitats likely to result in the all-too-near future from
anthropogenic climate change, forms of rewilding such

as assisted colonization and ecological substitution also challenge what Rob Nixon calls the eco-parochialism of conservation. All too often, Nixon suggests, conservation has hinged on hermetically sealed definitions of ecosystems, downplaying the permeable boundaries of bioregions and ignoring the spatial networks and exchanges that have always linked diverse natural spaces.[139] By challenging such inherently exclusionary ideologies of environmental space, rewilding offers an important alternative to the potentially xenophobic spatial foundations of environmentalism.

How much lost environmental time does rewilding propose to redeem? The reintroduction of wolves to Yellowstone restores the park to a period before European settler colonialism. Many rewilders, however, are unwilling to stop there. Radical conservationists have begun arguing for a *Pleistocene rewilding*. By reintroducing the large vertebrate species killed off as *Homo sapiens* spread around the world, this rewilding would, they argue, return ecosystems to an equilibrium state in place before the arrival of mankind.

Extinct megafauna putatively played an essential role in maintaining ecosystems through browsing, seed-dispersal, and predation. Their loss radically impoverished and unbalanced the environment. The rewilders challenge the idea, central to wilderness preservation in the US, that the state of wilderness encountered during European colonization of the Americas was pristine. What is natural for the Pleistocene rewilders is not the pre-European, but the pre-human. Pleistocene rewilders expose the notion of wilderness that has structured conservation efforts in the US and elsewhere as a myth, but only in order to push their notion of wilderness utopia back in time. If Native Americans had to be removed from the land in order to establish hallowed national spaces of wilderness like Yellowstone and Yosemite, the contemporary rewilding movement proposes an even more romantic construction of a sublime—and people-less—wilderness.[140]

Thus far, the most concrete embodiment of the idea is the Russian scientist Sergey Zimov's Pleistocene Park, established in 1989 in the remote wastes of Siberia. Zimov has stocked the massive park with large, extremely cold-resistant herbivores such as Yakutian horses, reindeer, moose, and musk oxen, as well as carnivores like foxes, bears, and wolves. He argues that these animals will help return the region from its current, relatively barren tundra state to what he calls "the mammoth ecosystem"—a grassland environment that preceded the Holocene extinctions.[141] Recreation of this ecosystem promises not just to revive lost biodiversity but, Zimov suggests, may also stave off climate change by preventing the massive release of carbon from melting Siberia permafrost. A corresponding movement for a Pleistocene rewilding of the US advocates the introduction of African and Asian elephants, lions, and cheetahs to parks set up on the Great Plains, filling the niche once occupied by mammoths, giant sloths, and other extinct megafauna, with similar benefits for biodiversity, climate change mitigation, and, not incidentally, ecotourism.[142] Challenging fundamental ideas about the conservation of unique ecosystems, rewilding advocates accept the most promiscuously anachronistic and anatopistic mixing of flora and fauna. At bottom, however, they are driven by the same nostalgia for a pristine natural world that has always animated wilderness philosophy. In focusing almost exclusively on Holocene extinctions, rewilders obscure the pivotal role of capitalism in global ecocide and ignore the violent and unequal histories of colonialism and imperialism that have ripped apart the planet.

Pleistocene rewilding is, however, hardly the most radical response to the extinction crisis. *De-extinction* promises to wind evolutionary time backwards even more vertiginously than rewilding. Efforts to use traditional methods of back-breeding to restore an approximation of

An ancestor of today's cattle, the Aurochs is the prized target of contemporary back-breeding programs. Credit: Pinpin via Wikimedia Commons.

extinct species have been around since the Heck brothers attempted to revive the aurochs, the large predecessor of all the world's cows, in Germany during the 1920s.

Today, the Tauros Programme is attempting to recreate the aurochs using similar back-breeding techniques, now guided by specific knowledge about the aurochs' genome drawn from molecular biology and genetics. The goal is to release the Tauros, a bovine breed that proponents claim will be indistinguishable from the aurochs, into European rewilded areas such as Holland's Oostvaardersplassen by 2020.[143]

But such efforts are conservative in comparison to the goals of de-extinction advocates who embrace the full potential of genomic technologies to resurrect extinct species like the woolly mammoth. An extinct animal has already been brought back to life: in 2000, a Franco-Spanish team transferred the nucleus of a skin cell from the world's last Pyrenean ibex, which was found dead in northern Spain earlier that year, into the egg cell of a domestic goat, implanting the cell into a surrogate mother in a process called interspecies nuclear transfer cloning.[144] Although the baby ibex died shortly after birth, the experiment showed that an extinct species could be brought back to life.

For scientists like George Church, inventor of the MAGE (multiplex automated genomic engineering) technology, synthetic biology promises nothing short of the resurrection of any extinct species whose genome is known or can be reconstructed from fossil remains. Key to this process is the conceptualization of animal species as bundles of genetic information, sequences of letters that can be stored on a computer. Animals (and humans, for that matter) are nothing more than genetic code, in this view, easily transposed into computer code.[145] Once the genomic information of a particular species is recovered or decoded, the problem simply becomes converting that information into strings of nucleotides that constitute the genes and genomes of the animal.[146] Church's MAGE technology massively accelerates the techniques of genetic engineering, allowing scientists to take the intact genome of one animal, say, an elephant,

and rejigger it using another animal's genome, a
mammoth for example, as a template, thereby creating a
new genome that can be used to clone a live mammoth
into existence using the same transfer cloning process
used to bring back the extinct Spanish ibex. Church,
with the aid of organizations such as *Revive and Restore*,
which bills itself as coordinating efforts at "genomic
conservation," hopes to resurrect not just the mammoth
but also the passenger pigeon, the Caribbean monk seal,
the golden lion tamarin, and many other now-extinct
animal species.[147]

If rewilders aim to restore lost environmental time,
synthetic biology promises, in Church's words, "to permit
us to replay scenes from our evolutionary past and to
take evolution to places where it has never gone."[148] De-
extinction is thus intimately tied to regenesis. MAGE has
in fact been nicknamed "the evolution machine" since
it can carry out the equivalent of millions of years of
genetic mutations within minutes. The biblical resonance
in the notion of regenesis is not ironic. Synthetic biology,
proponents argue, will quite literally make us into gods
by allowing us not just to resurrect extinct life forms
but also to engineer new life according to our needs.[149]
Not surprisingly, this promise has fired the public
imagination and attracted a fair amount of venture

capital to organizations like Revive and Restore. But even the most ardent proponents of de-extinction admit that many problems remain to be resolved. Biologists are, for instance, far better at manipulating genomes than at rewilding landscapes.[150] Furthermore, does it really make sense, many conservation biologists wonder, to spend huge amounts of time and capital to resurrect an extinct species such as the passenger pigeon when the threats that were responsible for its extinction—such as habitat destruction—have only intensified? We may be able to resurrect a few individual specimens of an extinct species, but will we not cruelly doom them to become extinct once again if we place them back in denuded habitats?

De-extinction offers a seductive but dangerously deluding techno-fix for an environmental crisis generated by the systemic contradictions of capitalism. It is not simply that de-extinction draws attention—and economic resources—away from other efforts to conserve biodiversity as it currently exists.[151] The fundamental problem with de-extinction is that it relies on the thoroughgoing manipulation and commodification of nature, and as such dovetails perfectly with *biocapitalism*. US lawyers have already begun arguing that revived species such as the mammoth would be "products of human ingenuity," and should therefore be eligible for patenting.[152] Species revival thus slots seamlessly into the neoliberal paradigms of research established by the Bayh-Dole Act of 1980, which legalized the patenting of scientific inquiry, as well as with the intellectual property agreements foisted on the world since the establishment of the World Trade Organization in the mid-1990s.[153]

De-extinction thus provides a mouth-watering opportunity for a new round of capital accumulation based on generating and acquiring intellectual property rights over living organisms. It is perhaps the most tangible and fully realized example of a shift that has been taking place since the 1980s, in which US petrochemical and pharmaceutical industries have reinvented themselves as purveyors of new, clean life sciences. Instead of generating (declining) profits through the mass-produced chemical fertilizers and pesticides of the Fordist era, agribusiness corporations like Monsanto have repositioned themselves to generate life itself by buying up biotech start-up companies. Capital is shifting, as Melinda Cooper observes, into "a new space of production—molecular biology—and into a new regime of accumulation, one that relies on financial investment to a much greater extent."[154] In this new post-mechanical age of production, the biological patent allows a company to own an organism's principle of generation, its genetic code, rather than owning the organism itself. Biological production is thereby transformed into capital's primary means for generating surplus value. Under this new regime of biocapitalism, living organisms are increasingly viewed, in the words of George Church and Ed Regis, as "programmable manufacturing systems."[155]

Biocapitalism is generated by and is deeply embedded in US imperialism. The massive investments in the life sciences that characterize this regime of accumulation are a product of the monetarist counterrevolution of 1979–1982, when the US introduced interest rate policies that channeled global financial flows into the dollar and US markets.[156] Since then, the US has financed its perpetually spiraling budget deficits through continuous inflows of capital. The result has been a form of capitalist delirium, which enables the US to operate—for a time—in utter disregard of economic and ecological limits. Yet US debt imperialism is based on the extraction of capital from vassal nations through the imposition of crippling structural adjustment policies by organizations like the International Monetary Fund and the World Bank.[157] Prostrated by debt, developing nations have been forced to sell off public assets and to open their economies to external capital penetration in a series of global enclosures of the commons. Ignoring these conditions of accumulation by dispossession, however, the ideologues of biocapitalism draw on the work of scientists such as Ilya Prigogine, whose *Order Out of Chaos* challenged the notions of limits inherent in the second law of thermodynamics by arguing that all of nature obeys the laws of self-organization and increasing complexity that characterize biological processes and systems.[158] Like life itself, the economy, neoliberals under the sway of this biocapitalist paradigm came to argue, is characterized by a process of continuous, self-regulating *autopoiesis* or self-engendering.[159] And again, like life, capitalism is said to be characterized by a series of catastrophic crises that ultimately generate new forms of complexity, as do mass extinction events in evolutionary history.

These neoliberal ideologies have come to permeate conservation to such an extent that discussions of biodiversity have become the site for the elaboration of what might be called *disaster biocapitalism*. Just as the disaster capitalism described by Naomi Klein seizes on political calamities to further its accumulative aims, this disaster biocapitalism takes the extinction crisis as an opportunity to ratchet up the commodification of life itself.[160] At the UN Climate Conference in 2007, for example, the UN and the World Bank announced the *Reducing Emissions from Deforestation and Forest Degradation* (REDD) scheme, which pays countries of the global South to reduce their deforestation and protect their existing forests. The carbon stored in these forests can then be quantified and sold to polluting industries in the global North, who can buy this stored carbon in order to "offset" rather than reduce their own polluting emissions. REDD was launched without input from Indigenous Peoples and other forest-dependent communities, and has already been linked to many land grabs and human rights violations.[161] All too often, local land stewards are represented in corporation-controlled international agreements like REDD as destroyers of biodiversity, and are consequently subjected to forced removal so that ecosystems can be privatized and reengineered as income-generating commodities to be sold on global capital markets.

Building on the REDD paradigm, the UN Convention on Biological Diversity in 2008 launched its own model for marketing "environmental services" through the Business and Biodiversity Initiative, which includes mechanisms for offsets and for the creation of "natural capital."[162] Within such schemes, the environmental commons of the global South, the planet's tropical forests and oceans and the myriad creatures who inhabit them, become a source of natural capital that can be quantified and traded on global markets. Biodiversity is thereby transmuted into a generator of offset credits that allow polluting corporations and governments to continue their ecological mayhem. Some of the world's most prominent conservation-based environmental NGOs have signed on to this disaster biocapitalism, including Conservation International, the Worldwide Fund for Nature, the Nature Conservancy, and the Environmental Defense Fund.[163] As we have seen, many of these conservation organizations are intensifying the social impact of the environmental crisis by encouraging states in the global South to evict Indigenous people, who are deemed incapable of managing their land, from conservation areas, creating a new category of "conservation refugees."[164]

7: RADICAL CONSERVATION

The philosophy of 'in the long run we are all dead' has guided economic development in the First and Third Worlds, in both socialist and capitalist countries. These processes of development have brought, in some areas and for some people, a genuine and substantial increase in human welfare. But they have also been marked by a profound insensitivity to the environment, a callous disregard for the needs of generations to come... It is what we know as the 'global green movement' that has most insistently moved people and governments beyond this crippling shortsightedness, by struggling for a world where the tiger shall still roam the forests of the Sunderbans and the lion stalk majestically across the African plain, where the harvest of nature may be more justly distributed across the members of the human species, where our children might more freely drink the water of our rivers and breathe the air of our cities.

—Ramachandra Guha, *Environmentalism: A Global History*

If mainstream environmentalism has been coopted by such neoliberal policies, what would a radical anti-capitalist conservation movement look like? It would begin from the understanding that the extinction crisis is at once an environmental issue *and* a social justice issue, one that is linked to long histories of capitalist domination over specific people, animals, and plants. The extinction crisis needs to be seen as an essential element in contemporary struggles against accumulation by dispossession. This crisis, in other words, ought to be a key issue in the fight for climate justice. If techno-fixes such as de-extinction facilitate new rounds of biocapitalist accumulation, an anti-capitalist movement against extinction must be framed in terms of a refusal to turn land, people, flora, and fauna into commodities. We must reject capitalist biopiracy and imperialist enclosure of the global commons, particularly when they cloak themselves in arguments about preserving biodiversity. Forums for enclosure such as the UNFCCC's Business and Biodiversity Initiative must be recognized for what they are and shut down. Most of all, an anti-capitalist conservation movement must challenge the privatization of the genome as a form of intellectual property, to be turned into an organic factory for the benefit of global elites. Synthetic biology should be regulated.[165] The genomic information of plants, animals, and human beings is the common wealth of the planet, and all efforts to make use of this environmental commons must be framed around principles of equality, solidarity, and environmental and climate justice.

Even well-meaning efforts to address extinction such as rewilding need to be challenged if they are not founded on considerations of globally redistributive climate justice. All too often rewilding schemes focus exclusively on wealthy areas of the planet. For instance, George Monbiot's "Manifesto for Rewilding the World" speaks exclusively of European rewilding schemes, and concludes by asking why Europe should not have a Serengeti or two.[166] This begs the question of what responsibility Europe has for Tanzania's Serengeti Park itself, as well as other wilderness areas in the global South. The record in this regard is deplorable. In 2013, for instance, Ecuador abandoned its Yasuni-ITT Initiative, which would have led to a moratorium on oil exploration in the Yasuni National Park, a UN biosphere reserve, in exchange for payments (by rich countries) of half the revenue drilling in the park was expected to generate.[167] The trust fund set up to manage the initiative received only a tiny fraction of the funds sought by Ecuador. How can one enthusiastically endorse rewilding in the global North when there is so little evidence of concrete determination to preserve existing biodiversity in the South? Moreover, if rewilding is seen as a way of saving charismatic African megafauna like the elephant from destruction by importing them to the badlands of Western Europe or North America, it will all too easily become a latter-day form of imperial ecology, creating glorified zoos stocked with purloined African and Asian wildlife.[168] Finally, rewilding makes strong arguments about the pivotal role of keystone species, but, in so doing, tends to reproduce the traditional bias in Western conservation efforts towards the large, the beautiful, and the charismatic. It is not a solution for the vast majority of flora and fauna threatened with extinction today.

An anti-capitalist conservation movement must not only be aware of histories of colonial expropriation of flora and fauna, but should focus on ways of fighting such forms of exploitation today. Wildlife in parks such as the Serengeti was revived following centuries of European colonial big-game hunting of native animals. Today, well-armed poachers again threaten megafauna in the world's remaining biodiversity hotspots. While the poachers tend to send their culls of elephant tusks and rhino horns mainly to foreign markets, in most cases their weapons come from decades of proxy battles during the Cold War. Moreover, African states are often unable to challenge these poachers as a result of IMF- and World Bank-administered structural adjustment policies that have left countries in the global

South on the brink of collapse. Efforts to deal with the extinction crisis cannot focus on rewilding the global North alone, nor should they focus exclusively on interdiction of the global traffic in wildlife. An anti-capitalist movement against extinction must also address the fundamental economic and political inequalities that drive the slaughter of megafauna. The extinction crisis should be framed in the context of a new wave of extractivism that is denuding many poor nations, shunting their minerals, flora, and fauna to consumer markets in industrialized nations. This new extractivism should be seen for what it is: a fresh wave of imperialism that is decimating poorer nations by removing the biological foundation of their collective future.[169]

What would be the shape and fundamental goals of an expansive anti-capitalist movement against extinction and for environmental justice? It would have to commence with open recognition by the developed nations of the long history of ecocide charted in this book. Such an admission would lead to a consequent recognition of the biodiversity debt owed by the wealthy nations of the global North to the South. Building on the demands articulated by the climate justice movement, the anti-capitalist conservation movement must demand the repayment of this biodiversity debt.[170] How would this repayment take place? As REDD demonstrates, states in the global South cannot always be counted on to disburse funds received from the North in a just manner; indeed, at present they collude all too often with resource-exploiting corporations by displacing genuine land stewards such as Indigenous and forest-dwelling peoples. The climate justice movement's call for a universal guaranteed income for inhabitants of nations who are owed climate debt should serve as a model here. Why not begin a model initiative for such a carbon and biodiversity-based guaranteed income program in the planet's biodiversity hotspots? Of the twenty-five terrestrial biodiversity hotspots, fifteen are covered primarily by tropical rainforests, and consequently are also key sites for the absorption of carbon pollution. These threatened ecosystems include the moist tropical woodlands of Brazil's Atlantic coast, southern Mexico with Central America, the tropical Andes, the Greater Antilles, West Africa, Madagascar, the Western Ghats of India, Indo-Burma, Indonesia, the Philippines, and New Caledonia. They make up only 1.4% of the Earth's surface, and yet, according to E.O. Wilson, these regions are "the exclusive homes of 44% of the world's plant species and more than a third of all species of birds, mammals, reptiles, and amphibians."[171] All of these areas are under heavy assault from the forces of enclosure and ecocide. A universal guaranteed income for the inhabitants of these hotspots would create a genuine counterweight to the attractions of poaching, and would entitle the Indigenous and forest-dwelling peoples who make these zones of rich biodiversity their homes with the economic and political power to push their governments to implement significant conservation measures.

Where would the capital for such a guaranteed income program for biodiversity hotspots come from? There is certainly no shortage of assets. As Andrew Sayer has argued, the 1% have accumulated their increasingly massive share of global wealth by siphoning off collectively produced surpluses not through hard work but through financial machinations such as dividends, capital gains, interests, and rent, much of which is then hidden in tax havens.[172] Indeed, if we consider the massive upward transfer of global wealth that has taken place over the last half century, it would be fair to say that never before was so much owed by so few to so many. One way to claw back some of this common wealth would be through a financial transactions tax of the kind proposed by James Tobin. Such a Robin Hood tax, of even only a very small percentage of the speculative global capital flows that enrich the 1%, would generate billions of dollars to help people conserve hotspots of global biodiversity. Such funds could also be devoted to ramping up renewable energy-generating infrastructures in both the rich and the developing countries.

Yet a universal guaranteed income in recognition of biodiversity debt should not be a replacement for existing conservation programs. Instead, such a measure should be seen as an effort to inject an awareness of environmental and climate justice into debates around the extinction crisis. Biodiversity debt would thus augment existing conservation programs while militating against the creation of conservation refugees. In addition, rewilding and de-extinction, despite their significant flaws, may have a place in an anti-capitalist conservation movement, but only if they are reframed in terms of the history of ecocide. Rewilding, for instance, should not be undertaken in the global North without a commensurate pledge of economic assistance for conservation and rewilding of areas in the global South, whose present depleted state is often a direct product of the North's extractive industries, from plantation slavery to the latest round of land grabs. Similarly, de-extinction may be employed judiciously, for example to reintroduce extinct versions of genes into species that have lost a dangerous amount of genetic diversity. Such efforts should, however, be designed to conserve existing biodiversity, particularly in endangered hotspots, rather than to resurrect extinct charismatic megafauna from the grave. Any and all such efforts to work against extinction should be undertaken as acts of environmental solidarity on the part of the peoples of the global North with the true stewards of the planet's biodiversity, the people of the global South. Only in this way can the struggle against extinction help promote not simply forgiveness and reconciliation, but also survival after five hundred years of colonial and imperial ecocide.

The struggle to preserve global biodiversity must be seen as an integral part of a broader fight to challenge an economic and social system based on feckless, suicidal expansion. If, as we have seen, capitalism is based on ceaseless compound growth that is destroying ecosystems the world over, the goal in the rich nations of the global North must be to overturn our present expansionary system by fostering *de-growth*. Most importantly, nations that have benefited from burning fossil fuels must radically cut their carbon emissions in order to stem the lurch towards runaway climate chaos that endangers the vast majority of current terrestrial forms of life. Rather than false and impractical solutions such as the carbon trading and geo-engineering schemes championed by advocates of neoliberal responses to the climate crisis, anti-capitalists should fight for some version of the contraction and convergence approach proposed by the Global Commons Institute.[173] This proposal is based on moving towards a situation in which all nations have the same level of emissions per person (convergence) while contracting them to a level that is sustainable (contraction). A country such as the United States, which has only 5% of the global population, would be allowed no more than 5% of globally sustainable emissions. Such a move would represent a dramatic anti-imperialist shift since the US is at present responsible for 25% of carbon emissions.

The powerful individuals and corporations that control nations like the US are not likely to accept such revolutionary curtailments of the wasteful system that supports them without a struggle. Already there is abundant evidence that they would sooner destroy the planet than let even a modicum of their power slip. Massive fossil fuel corporations such as Exxon, for example, have funded climate change denialism for the past quarter century despite abundant evidence *from their own scientists* that burning fossil fuels was creating unsustainable environmental conditions.[174] Such behavior should be seen frankly for what it is: a crime against humanity. We should not expect to negotiate with such destructive entities. Their assets should be seized. Most of these assets, in the form of fossil fuel reserves, cannot be used anyway if we are to avert environmental catastrophe. What remains of these assets should be used to fund a rapid, managed reduction in carbon emissions and a transition to renewable energy generation. These steps should be part of a broader program to transform the current, unsustainable capitalist system that dominates the world into steady state societies founded on principles of equality and environmental justice.

At present, such revolutionary measures seem completely impractical since most of the media, the political parties, and the repressive power of the state are in the hands of the plutocrats. Yet now, more than ever, we cannot let the present state of affairs determine our horizon of hope and sense of possibility. The terminal crisis of capitalism is no longer a prospect—it is a reality that is breaking across the planet like a series of ferocious interconnected storms. Science tells us that this unprecedented climate turbulence will first wash over tropical, postcolonial nations, where decades of structural adjustment have weakened infrastructure, fed urban destitution, and decimated collective solidarities.[175] Already we are seeing climate change-catalyzed conflicts such as the war in Syria devastate entire societies, generating millions of refugees, thousands of whom have been left in limbo by the refusal of European nations to offer safe harbor.[176] Yet while the global South will be hit first and hardest, the coming waves of climate chaos will wash across the entire globe. As Christian Parenti has argued, there are no safe harbors from this gathering storm.[177] Ironically, continuing with business as usual is now a recipe for increasingly catastrophic disruption of the basic climatic conditions humanity has enjoyed since the Neolithic Revolution. Inaction is now a recipe for dissolution. Simply in order to retain an environment conducive to the continued existence of our fellow animals, plants, and humans, then, we must transform the root conditions of the climate crisis: the unsustainable capitalist system that is driving the sixth extinction. In sum, the only true conservation is a radical conservation.

8: CONCLUSION

The pika is a small, rather cute mammal that looks a bit like a hamster. Pikas live on the rugged slopes of mountain ranges in eastern Asia, the Middle East, and North America. Researchers report that extinction rates for the American pika have increased nearly five-fold in the last decade.[178] Since they depend on cool, high-mountain habitats to survive, pikas have been coping with the higher temperatures caused by climate change by moving up mountain slopes at a rate that has increased eleven-fold over the last ten years. Pikas eventually arrive at the top of their mountains; at this point, they have nowhere left to go to escape global warming. Their desperate plight is a particularly poignant metaphor for the situation in which the animal and plant species of our planet as a whole increasingly find themselves. As the first mammal species to be directly endangered by climate change, pikas are a sentinel species, a warning of the intensification of already catastrophic rates of habitat destruction and extinction resulting from anthropogenic climate change.

Why should we care about the fate of the diminutive pika, or any other endangered species of plant or animal for that matter? Why bother about extinction? Aren't there many other crises, including environmental calamities such as city-destroying hurricanes, to worry about? These questions can be answered in purely utilitarian terms. Human beings depend on other species for our existence. The plants and animals that surround us synthesize the oxygen we breathe, consume the carbon dioxide we emit, produce the food we eat, maintain the fertility of the soil, and return our bodies to the earth after we die.[179] Although many cultures recognize and celebrate this rich interdependency of species, the capitalist system that has come to dominate the world over the last five centuries is grounded in and thrives on dispossession. When viewed through the lens of extinction, it is an economic system and culture founded on a drive to annihilate everything in its path.

But there is a very different answer to the question of why we should bother about extinction. Each species and ecosystem contributes to the richness and beauty of life on Earth. Each is unique and, according to the increasingly influential doctrines of Earth jurisprudence or Wild Law, each is an integral part of the web of life and consequently has rights that must be recognized and revered.[180] Once a species or an ecosystem is destroyed, it is lost forever. The great wave of destruction that is the sixth extinction radically impoverishes not just the planet but humanity as well. It is an indication that something has gone fundamentally wrong with us. Some might suggest that human beings have been cursed with the capacity to destroy other species wholesale for many millennia. In her influential book on extinction, for example, Elizabeth Kolbert writes that it is our creativity as a species that endangers, but also may save the planet. "As soon as humans started using signs and symbols to represent the natural world," she writes, "they pushed beyond the limits of that world."[181] There is certainly some ground for this assertion: as we have seen, language allowed *Homo sapien*s to organize themselves into lethal hunting bands capable of unleashing a worldwide wave of megafauna extinction in the late Pleistocene era. Yet since then, many human cultures have learned to live in relative harmony with the flora and fauna that surround them. More importantly, the extinctions of past millennia pale in comparison with the decimation of global wildlife unleashed by capitalism during the modern era. Understanding that capitalism is responsible for the lion's share of the sixth extinction helps us avoid the deeply dystopian idea that human beings are innately destructive of the natural world.

An anti-capitalist perspective also prevents us from attributing ecocide to humanity as a whole. As we have seen, capitalism has unleashed waves of enclosure, imperialism, warfare, and ecocide over the last five hundred years that have benefitted a very small segment of humanity while displacing, immiserating, enslaving, and destroying countless numbers of people, animals, and plants. Everyone is not equally responsible for the destruction of nature, despite Kolbert's suggestion that "if you want to think about why humans are so dangerous to other species, you can picture a poacher in Africa carrying an AK-47 or a logger in the Amazon gripping an ax, or, better still, you can picture yourself, holding a book in your lap."[182] Such a sweeping indictment of an undifferentiated humanity is both historically inaccurate and politically disempowering. Such a perspective offers us no understanding of the structural forces that generate exploitation and ecocide, no sense of how such forces may push the vulnerable to behave in ways that are antithetical to their long-term interest, and no conception of how people in the relatively affluent global North might act in solidarity with those whom Frantz Fanon called "the wretched of the earth." Such a perspective is truly hopeless.

It has been said that it is easier to imagine the end of the world than to envisage the overthrow of capitalism.[183] I would respond to this aphorism from dark times that it is easier to imagine the end of capitalism than it is to articulate any other genuine solution to the extinction crisis. If capitalism is the ultimate cause and prime engine of the extinction crisis, surely we can only conclude that we may find hope in challenging its baleful power with all means at our disposal. Capitalism is not eternal; it is a specific economic system grounded in a set of historically particular economic arrangements and social values. It came onto the world stage relatively recently, and, one way or another, it will eventually make an exit. The question for us, then, is what kind of end we wish to make. Thinking in anti-capitalist terms can be liberating, triggering myriad constructive projects and emancipatory prospects. Indeed, as Naomi Klein has recently argued, the climate crisis is already stirring up many novel experiments and exciting visions for a new society.[184] But Klein's point is an even more fundamental one: climate science, she points out, has made it blindingly clear that our economic system is destroying the planetary life support systems upon which we depend. Climate change therefore makes it imperative that we discuss radical transformations in capitalist social relations, a topic that has been largely taboo for the last two decades.

The extinction crisis makes the urgency of the transformation Klein alludes to even more palpable. After all, increasing atmospheric carbon concentrations remain relatively abstract for most people on the planet. In contrast, the wave of extinction that is decimating plants and animals around the planet strikes at the most intimate and potent of human faculties: our ability to imagine. The power of human dreams has historically been closely tied to the generative multiformity of the plant and animal life that surrounds us.[185] Even in the "advanced" capitalist cultures, we encourage our children to learn basic forms of empathy and imagination by giving them toy animals and reading them stories like *The Tale of Peter Rabbit*. We have always used animals and plants to symbolize our most intimate fears, our hopes, and even our greatest loves. As capitalism tears increasingly gaping holes in the beautiful web of life of which we are a part, our capacity to dream, to imagine different, more manifold worlds is radically impoverished. Every species that is consigned to oblivion is a grave loss to the planet in general and a serious threat to the many people whose lives are intertwined with that species. In addition, however, such losses are the most concrete possible testimony to the ecocidal character of capitalism. In the face of such an irredeemably rapacious and ultimately impoverishing system, we must insist on the human capacity to dream and to build a more just, more biologically diverse world.

BIBLIOGRAPHY

Anker, Peder. *Imperial Ecology*. Cambridge, MA: Harvard
    UniversityPress, 2009.

Armiero, Marco, *Wasteocene: Stories from the Global Dump*
    (New York: Cambridge University Press, 2021).

Arneil, Barbara. *John Locke and America: The Defense of English
    Colonialism*. New York: Oxford University Press, 1996.

Beever, Erik A., Chris Ray, Jenifer L. Wilkening, Peter F.
    Brussard, Philip W. Mote, "Contemporary climate change
    alters the pace and drivers of extinction" *Global Change
    Biology*, 2011; DOI: 10.1111/j.1365-2486.2010.02389.x.

Bhandar, Brenna, Colonial Lives of Property: Law, Land, and Racial
    Regimes of Ownership (Durham, NC: Duke University Press,
    2018).

Brashares, Justin S. et al. "Wildlife Decline and Social Conflict,"
    Science 345.6195 (25 July 2014): 376–378.

Broswimmer, Franz J. *Ecocide: A Short History of the Mass
    Extinction of Species*. New York: Pluto, 2002.

Carlin, Dr. Norman, Ilan Wurman, and Tamara Zakim, "How
    To Permit Your Mammoth: Some Legal Implications of
    'De-Extinction," *Stanford Environmental Law Journal* 33.3
    (January 2014), https://journals.law.stanford.edu/stanford-

environmental-law-journal-selj/print/volume-33/number-1/
how-permit-your-mammoth-some-legal-implications-de-
extinction.

Carrington, Damian. "Earth Has Lost Half Its Wildlife in the
    Past Forty Years, WWF Says," *The Guardian* (30 September
    2014), http://www.theguardian.com/environment/2014/
    sep/29/earth-lost-50-wildlife-in-40-years-wwf.

Chakrabarty, Dipesh. "The Climate of History," *Critical Inquiry*
    35 (Winter 2009), 197–222.

Christy, Brian. "Blood Ivory," *National Geographic*
    (October 2012), Accessed 5 August 2014, http://ngm.
    nationalgeographic
    .com/2012/10/ivory/christy-text.

Church, George and Ed Regis. *Regenesis: How Synthetic Biology
    Will Reinvent Nature and Ourselves*. New York: Basic
    Books, 2012.

Cohen, Mark Nathan. *The Food Crisis in Prehistory:
    Overpopulation and the Origins of Agriculture*. New Haven,
    CT: Yale University Press, 1977.

Columbus, Christopher. *Select Letters of Christopher Columbus*,
    R.S. Major, trans. and ed.. London: Hakluyt Society, 1870.

Cooper, Melinda. *Life as Surplus: Biotechnology and Capitalism
    in the Neoliberal Era*. Seattle, WA: University of Washington
    Press, 2008.

Counsell, Simon, "Anatomy of a 'Nature-Based Solution': Total
    Oil, 40,000 Hecatres of Disappearing African Savannah,
    Emmanuel Macron, Norwegian and French 'Aid' to an
    Election-Rigging Dictator, Trees to Burn, Secret Contracts,

and Dumbstruck Conservationists," 16 Apr 2021, https://
redd-monitor.org/2021/04/16/anatomy-of-a-nature-based-
solution-total-oil-40000-hectares-of-disappearing-african-
savannah-emmanuel-macron-norwegian-and-french-aid-to-
an-election-rigging-dictator-trees/

Cronon, William. "The Trouble with Wilderness; or, Getting
Back to the Wrong Nature," in William Cronon, ed.,
*Uncommon Ground: Rethinking the Human Place in
Nature*. New York: W. W. Norton & Co., 1995, 69–90.

Crosby, Alfred. *The Columbian Exchange: Biological and
Cultural Consequences of 1492*. New York: Praeger, 2003.

Crutzen, Paul. "The Geology of Mankind," *Nature* 415.6867
(2002), 23.

Cullinan, Cormac. *Wild Law: A Manifesto for Earth Justice*.
New York: Chelsea Green, 2011.

Dell'Amore, Christine. "Beloved African Elephant Killed for
Ivory," *National Geographic* (16 June 2014), Accessed
5 August 2014, http://news.nationalgeographic.com/
news/2014/06/140616-elephants-tusker-satao-poachers-
killed-animals-africa-science/.

Demos, T.J., "Climate Control: From Climate Emergency
to Climate Emergence," e-flux journal 104 (Nov 2019),
https://www.e-flux.com/journal/104/299286/climate-
control-from-emergency-to-emergence/

Dirzo, Rodolfo. "Defaunation in the Anthropocene," *Science*
345.6195 (25 July 2014): 401–406.

Donlan, Josh. "De-extinction in a Crisis Discipline," *Frontiers of
Biogeography* 6.1 (2014), 25–28.

Dowie, Mark, Conservation Refugees: The Hundred Year Conflict Between Global Conservation and Native Peoples (Cambridge, MA: MIT Press, 2009).

———, "Conservation Refugees," *Orion Magazine*, https://orionmagazine.org/article/conservation-refugees/

——— "Lions and Cheetahs and Elephants, Oh My!" *Slate* (August 18, 2005), Accessed 5 August 2014, http://www.slate.com/articles/health_and_science/science/2005/08/lions_and_cheetahs_and_elephants_oh_my.html.

Dunbar-Ortiz, Roxanne. *An Indigenous People's History of the United States*. New York: Beacon, 2014.

Fanon, Frantz, *The Wretched of the Earth* (New York: Grove, 2005).

Federici, Silvia. *Caliban and the Witch: Women, the Body, and Primitive Accumulation*. New York: Autonomedia, 2004.

Ferguson, Brian R. "Ten Points on War," *Social Analysis*, 52.2 (Summer 2008), 32–49.

Fischer, Steven Roger. *History of Writing*. New York: Reaktion Books, 2004.

Gadgil, Mahdav and Ramachandra Guha, *This Fissured Land: An Ecological History of India* (New York: Oxford University Press, 2012).

IUCN, "IUCN Congress Closes With Call for a Post-Pandemic Recovery Based in Nature" (10 Sept 2021), https://www.iucn.org/news/secretariat/202109/iucn-congress-closes-call-a-post-pandemic-recovery-based-nature

———, "Motion 101," https://www.iucncongress2020.org/motion/101

Global Justice Ecology Project, *The Green Shock Doctrine* (12 May 2014), Accessed 5 August 2014, http://globaljusticeecology.org/green-shock-doctrine/.

Grainger, Sally, ed., *Apicius: A Critical Edition*. New York: Prospect Books, 2006.

Grove, Richard. *Green Imperialism: Colonial Expansion, Tropical Island Edens, and the Origins of Environmentalism, 1600–1860*. New York: Cambridge University Press, 1996.

Guha, Ranajit and Juan Martinez-Alier, *Varieties of Environmentalism: Essays North and South*. London: Earthscan, 1997.

Haraway, Donna. *When Species Meet*. Minneapolis: University of Minnesota Press, 2007.

Hardt, Michael and Antonio Negri. *Commonwealth*. Cambridge, MA: Belknap Press, 2011.

Harrison, Robert Pogue. *Forests: The Shadow of Civilization*. Chicago, IL: University of Chicago Press, 1992.

Harvey, David. *Seventeen Contradictions and the End of Capitalism*. New York: Oxford University Press, 2014.

—— *The Enigma of Capital*. New York: Oxford University Press, 2010.

—— *The New Imperialism*. New York: Oxford University Press, 2003.

Heise, Ursula. "Lost Dogs, Last Birds, and Listed Species: Cultures of Extinction," *Configurations* 18.1–2 (Winter 2010), 49–72.

Hughes, J. Donald. *Environmental Problems of the Greeks and Romans: Ecology in the Ancient Mediterranean*. Baltimore, MD: Johns Hopkins University Press, 2014.

——— "Ripples in Clio's Pond: Rome's Decline and Fall: Ecological Mistakes?" *Capitalism, Nature, Socialism* 8.2 (June 1997), 117–21.

Jameson, Fredric. "Future City," New Left Review 21 (May-June 2003), Accessed 5 August 2014, http://newleftreview.org/II/21/fredric-jameson-future-city.

Kashwan, Prakash, Rosaleen Duffy, Francise Massé, Adeniyi Asiyanbi, Esther Marijnen, "From Racialized Neocolonial Global Conservation to an Inclusive and Regenerative Conservation," *Environmental* 63.4 (2021), 4–19.

Klein, Naomi. *The Shock Doctrine: The Rise of Disaster Capitalism*. New York: Picador, 2008.

——— *This Changes Everything: Capitalism vs. The Climate*. New York: Simon and Schuster, 2014.

Kolbert, Elizabeth. "Recall of the Wild: The Quest to Engineer a World Before Humans," *The New Yorker* (24 December 2012), Accessed 5 August 2014, http://www.newyorker.com/magazine/2012/12/24/recall-of-the-wild.

——— "Save the Elephants," *New Yorker* (7 July 2014), Accessed 5 August 2014, http://www.newyorker.com/magazine/2014/07/07/save-the-elephants.

——— *The Sixth Extinction: An Unnatural History*. New York: Henry Holt, 2014.

Kovel, Joel. *The Enemy of Nature: The End of Capitalism or the End of the World*. New York: Zed, 2007.

Lascher, John. "If You Plant Different Trees in the Forest, Is It Still the Same Forest?" *The Guardian* (19 October 2014), Accessed 5 August 2014, http://www.theguardian.com/vital-signs/2014/oct/19/-sp-forests-nature-conservancy-climate-change-adaptation-minnesota-north-woods.

Liberti, Stefano. *Land Grabbing: Journeys in the New Colonialism*. New York: Verso, 2014.

Lilley, Sasha, David McNally, and Eddie Yuen, *Catastrophism: The Apocalyptic Politics of Collapse and Rebirth*. New York: PM Press, 2012.

Locke, John. *Second Treatise on Government*, Chapter 5: Of Property. Accessed 5 August 2014, http://www.constitution.org/jl/2ndtr05.htm.

Longo, Fiore, "Personal Interview," 21 Dec 2021.

MacKinnon, J. B. *The Once and Future World: Nature As It Was, As It Is, As It Could Be*. New York: Houghton Mifflin Harcourt, 2013.

Martin, Paul S. *Twilight of the Mammoths: Ice Age Extinctions and the Rewilding of America*. Berkeley, CA: University of California Press, 2005.

Merchant, Carolyn, *The Death of Nature: Women, Ecology, and the Scientific Revolution*. New York: HarperOne, 1990.

Monbiot, George. "A Manifesto For Rewilding the World," Accessed 5 August 2014, http://www.monbiot.com/2013/05/27/a-manifesto-for-rewilding-the-world/.

——— *Feral: Searching for Enchantment on the Frontiers of Rewilding*. New York: Allen Lane, 2013.

Moore, Jason. "Anthropocene or Capitalocene?" Accessed
    9 August 2014, http://jasonwmoore.wordpress.com/2013/
    05/13/anthropocene-or-capitalocene/.

Nixon, Rob. "Environmentalism and Postcolonialism," in
    Ania Loomba and Suvir Kaul, eds., *Postcolonial Studies
    and Beyond*. Durham, NC: Duke University Press, 2005),
    233–51.

O'Connor, James. *Natural Causes: Essays in Ecological
    Marxism*. New York: Guilford Press, 1997.

"Our Land, Our Nature" Congress, https://www.
    ourlandournature.org/

Parenti, Christian. *Tropics of Chaos: Climate Change and the
    New Geography of Violence*. New York: Nation Books, 2012.

Parr, Adrian. *The Wrath of Capital: Neoliberalism and Climate
    Change Politics*. Columbia University Press, 2014.

Ponting, Clive. *A Green History of the World: The Environment
    and the Collapse of Great Civilizations*. NY: Penguin, 1991.

Prigogine, Ilya and Isabelle Stengers, *Order Out of Chaos: Man's
    New Dialogue With Nature*. New York: Bantam, 1984.

"Regulate Synthetic Biology Now: 194 Countries," *SynBio
    Watch* (20 October 2014), http://www.synbiowatch.
    org/2014/10/regulate-synthetic-biology-now-194-
    countries/.

Revkin, Andrew. "Confronting the Anthropocene," *New York
    Times* (11 May 2011), Accessed 9 August 2014, http://
    dotearth.blogs
    .nytimes.com/2011/05/11/confronting-the-anthropocene/?
    _php=true&_type=blogs&_r=0.

Rewilding Europe at http://www.rewildingeurope.com/.

Rich, Nathaniel. "The Mammoth Cometh," *New York Times Magazine* (27 February 2014), Accessed 9 August 2014, http://www.nytimes.com/2014/03/02/magazine/the-mammoth-cometh.html.

Richards, John F. *The World Hunt: An Environmental History of the Commodification of Animals.* Berkeley, CA: University of California Press, 2014.

Roberts, Neil. *The Holocene: An Environmental History.* New York: Basil Blackwell, 1992.

Ross, Andrew. *Creditocracy and the Case for Debt Refusal.* New York: OR Books, 2014.

Ross, Kristin, Communal Luxury: The Political Imaginary of the Paris Commune (New York: Verso, 2016).

Ruddiman, William F. "The anthropogenic greenhouse era began thousands of years ago". *Climatic Change* 61.3 (2003): 261–293.

Sanders, Barry. *The Green Zone: The Environmental Costs of Militarism.* Oakland, CA: AK Press, 2009.

Schleicher et.al., "Protecting Half of the Planet Would Directly Affect over One Billion People," *Nature Sustainability* 2 (Dec 2019), 1094-1096.

Scientific American Editors. "Why Efforts To Bring Extinct Species Back from the Dead Miss the Point," Scientific American 308.6 (14 May 2013), Accessed 9 August 2014, http://www.scientificamerican.com/article/why-efforts-bring-extinct-species-back-from-dead-miss-point/.

Seddon, Philip *et al.*, "Reversing Defaunation: Restoring Species in a Changing World," *Science* 345.6195 (2014): 406–412.

Shiva, Vandana. *Monocultures of the Mind: Biodiversity,
    Biotechnology, and Agriculture*. New Delhi: Zed Press, 1993.
⸻ *Stolen Harvest: The Hijacking of the Global Food Supply*.
    Boston, MA: South End Press, 2000.
⸻ *The Violence of the Green Revolution: Third World
    Agriculture, Ecology, and Politics*. New York: Zed Books,
    1992.

Shiva, Vandana and Ingunn Moser, eds., *Biopolitics: A Feminist and
    Ecological Reader on Biotechnology*. Atlantic Highlands, NJ: Zed,
    1995.

Smith, Adam, *The Wealth of Nations*. New York: Bantam Classic,
    2003.

Solnit, Rebecca. *Savage Dreams: A Journey in the Hidden Wars
    of the American West*. San Francisco: Sierra Club, 1994.

Thacker, Eugene. *The Global Genome: Biotechnology, Politics,
    and Culture*. Cambridge, MA: MIT Press, 2005.

Tollefson, Jeff, "'Global Deal for Nature' Fleshed Out with
    Specific Conservation Goals" *Nature* (19 Apr 2019), https://
    www.nature.com/articles/d41586-019-01253-z

United Nations, "Sustainable Development Goals - #12: Ensure
    Sustainable Production and Consumption Patterns,"
    https://www.un.org/sustainabledevelopment/sustainable-
    consumption-production/

The Rewilding Institute at http://rewilding.org/rewildit/.

van Dooren, Toom. *Flight Ways: Life and Loss at the Edge of
    Extinction*. New York: Columbia University Press, 2014.

Veltmeyer , Henry and James Petras, eds., *The New Extractivism: A
    Post-Neoliberal Development Model or Imperialism of the 21st
    Century*. New York: Zed Books, 2014.

Vidal, John. "How the Kalahari Bushmen and Other
    Tribespeople Are Being Evicted to Make Way for
    'Wilderness,'" *The Guardian* (15 November 2014), Accessed
    9 August 2014, http://www.theguardian.com/world/2014/
    nov/16/kalahari-bushmen-evicted-wilderness.

Vigneri, Sacha. "Vanishing Fauna," *Science* 345.6195 (25 July
    2014): 393–395.

Watts, Jonathan. "Ecuador Approves Yasuni National Park Oil
    Drilling in Amazon Forest," *The Guardian* (16 August 2013),
    Accessed 5 August 2014, http://www.theguardian.com/
    world/2013/aug/16/ecuador-approves-yasuni-amazon-oil-
    drilling.

Williams, Eric. *Capitalism and Slavery*. Charlotte, NC:
    University of North Carolina Press, 1994.

Wilson, Edward O. *The Future of Life*. New York: Knopf, 2004.

———, *Half-Earth: Our Planet's Fight for Life* (New York: Liveright,
    2017).

WWF, "Morges Manifesto," https://wwfeu.awsassets.panda.org/
    downloads/morgesmanifesto.pdf

WWF, *Living Planet Report 2014*, Accessed 5 August 2014,
    http://wwf.panda.org/about_our_earth/all_publications/
    living_planet_report/.

Zierler, David. *The Invention of Ecocide: Agent Orange, Vietnam,
    and the Scientists Who Changed the Way We Think About
    the Environment*. Athens, GA: University of Georgia Press,
    2001.

Zimov, Sergey. "Pleistocene Park: Return of the Mammoth's
    Ecosystem," *Science* 308.5723 (6 May 2005), 796–798.

ACKNOWLEDGEMENTS

My profound thanks to Anne McClintock and Rob Nixon for your enduringly inspirational work and presence in the world. I feel very fortunate to have been able to count the two of you as mentors and friends across the span of decades.

I am very grateful to the community at the Blue Mountain Center, where I spent an idyllic month writing and reflecting. It was at BMC that the idea for this project was born.

Thanks to Colleen Lye, an old friend who helped set me up with research privileges at Berkeley, where much of the writing of this book was done.

Massive thanks to my amazing research assistants Sarah Hildebrand and Stefano Morello for all your help with this project.

I am immensely grateful and deeply indebted to Eddie Yuen for our many conversations about capitalism and extinction. These conversations were germinal to everything written here. Any erroneous elements in the text are certainly my fault, but whatever felicities found their way into this book are surely linked to my discussions of extinction with Eddie.

Thanks to Colin Robinson of OR Books for his thoughtful editing and enthusiastic support for this book. I am also very grateful to the rest of the crew at OR for their work on the project.

Last of all, I thank Manijeh Moradian for unwavering love and wisdom. The time during which this book was written was one of immense happiness and growth, much of which I owe to you, azizam.

ENDNOTES

1. Caspar A. Hallmann et al., "More than 75 Percent Decline over 27 Years in Total Flying Insect Biomass in Protected Areas," *PLOS ONE* 12, no. 10 (October 18, 2017): e0185809, https://doi.org/10.1371/journal.pone.0185809.

2. Caspar et al., "More than 75 Percent Decline"

3. Brooke Jarvis, "The Insect Apocalypse Is Here," *The New York Times*, November 27, 2018, sec. Magazine, https://www.nytimes.com/2018/11/27/magazine/insect-apocalypse.html.

4. "'Insect Apocalypse' Poses Risk to All Life on Earth, Conservationists Warn," *The Guardian*, November 13, 2019, http://www.theguardian.com/environment/2019/nov/13/insect-apocalypse-poses-risk-to-all-life-on-earth-conservationists-warn.

5. "Two-Fifths of Plants at Risk of Extinction, Says Report," *BBC News*, September 30, 2020, sec. Science & Environment, https://www.bbc.com/news/science-environment-54344309.

6. "Humanity Has Wiped out 60% of Animal Populations since 1970, Report Finds," *The Guardian*, October 30, 2018, http://www.theguardian.com/environment/2018/oct/30/

humanity-wiped-out-animals-since-1970-major-report-finds.

7. "There's Nothing 'Mini' about the U.S. Plan to Unravel Europe's Precautionary Principle," accessed April 7, 2021, https://www.iatp.org/blog/202002/theres-nothing-mini-about-us-plan-unravel-europes-precautionary-principle.

8. Ivette Perfecto, John Vandermeer, and Angus Wright, *Nature's Matrix: Linking Agriculture, Biodiversity Conservation and Food Sovereignty* (New York: Routledge, 2019).

9. United Nations Collaborative Program on Reducing Emissions from Deforestation and forest Degradation in Developing Countries (REDD+), https://www.un-redd.org/

10. "Xapuri Declaration: 'We Reject Any Form of Climate Colonialism' | Indigenous Environmental Network," June 20, 2017, https://www.ienearth.org/xapuri-declaration-we-reject-any-form-of-climate-colonialism/.

11. See E.O. Wilson, *Half-Earth: Our Planet's Fight for Life* (New York: Liveright, 2016), as well as Troy Vettese, "To Freeze the Thames: Natural Geo-Engineering and Biodiversity," *New Left Review* 111 (May/June 2018).

12. "Indigenous Peoples Defend Earth's Biodiversity—but They're in Danger," *National Geographic,* (November 16, 2018), https://www.nationalgeographic.com/environment/article/can-indigenous-land-stewardship-protect-biodiversity-.

13. Rob Wallace, *Dead Epidemiologists: On the Origins of COVID-19* (New York: Monthly Review Press, 2020) and

Mike Davis, *The Monster Enters: COVID-19, Avian Flu, and the Plagues of Capitalism* (New York: O/R Books, 2020).

14. Christine Dell'Amore, "Beloved African Elephant Killed for Ivory," *National Geographic* (16 June 2014), Accessed 5 August 2014, http://news.nationalgeographic.com/news/2014 /06/140616-elephants-tusker-satao-poachers-killed-animals-africa-science/. See also Brian Christy, "Blood Ivory," *National Geographic* (October 2012), Accessed 5 August 2014, http://ngm.nationalgeographic.com/2012/10/ivory/ christy-text.

15. Elizabeth Kolbert, "Save the Elephants," *New Yorker* (7 July 2014), Accessed 5 August 2014, http://www.newyorker. com/magazine/2014/07/07/save-the-elephants.

16. Sacha Vignieri, "Vanishing Fauna," *Science* 345.6195 (25 July 2014): 393–395.

17. Edward O. Wilson, *The Future of Life* (New York: Knopf, 2004), 92.

18. Rudolfo Dirzo, "Defaunation in the Anthropocene," *Science* 345.6195 (25 July 2014): 401–406.

19. Dirzo, 401.

20. Wilson, 99.

21. Franz J. Broswimmer, *Ecocide: A Short History of the Mass Extinction of Species* (New York: Pluto, 2002), 1.

22. Elizabeth Kolbert, *The Sixth Extinction: An Unnatural History* (New York: Henry Holt, 2014), 167.

23. Dirzo, 401.

24. Wilson, 59.

25. Wilson, 59.

26. See, for example, Donna Haraway, *When Species Meet* (University of Minnesota Press, 2007), Ursula Heise, "Lost Dogs, Last Birds, and Listed Species: Cultures of Extinction," *Configurations* 18.1–2 (Winter 2010), 49–72 and Thom van Dooren, *Flight Ways: Life and Loss at the Edge of Extinction* (New York: Columbia University Press, 2014).

27. Justin S. Brashares et al., "Wildlife Decline and Social Conflict," *Science* 345.6195 (25 July 2014): 377.

28. See Brashares et al. for a critique of the "war on poachers," although they rather characteristically do not link this militarized response to poaching to the broader politics of violence that characterizes the "war on terror."

29. Christian Parenti, *Tropics of Chaos: Climate Change and the New Geography of Violence* (New York: Nation Books, 2012).

30. Michael Hardt and Antonio Negri, *Commonwealth* (Cambridge, MA: Belknap Press, 2011), viii.

31. Vandana Shiva, *Stolen Harvest: The Hijacking of the Global Food Supply* (Boston, MA: South End Press, 2000).

32. James O'Connor, *Natural Causes: Essays in Ecological Marxism* (New York: Guilford Press, 1997), 166.

33. David Harvey, *Seventeen Contradictions and the End of Capitalism* (New York: Oxford University Press, 2014), 222.

34. Vandana Shiva, *Monocultures of the Mind: Biodiversity, Biotechnology, and Agriculture* (New Delhi: Zed Press, 1993).

35. Harvey, *Seventeen Contradictions*, 254.

36. Sasha Lilley, David McNally, and Eddie Yuen, *Catastrophism: The Apocalyptic Politics of Collapse and Rebirth* (New York: PM Press, 2012).

37. Ranajit Guha and Juan Martinez-Alier, *Varieties of Environmentalism: Essays North and South* (London: Earthscan, 1997), 12.

38. Dirzo, 404.

39. Andrew Revkin, "Confronting the Anthropocene," *New York Times* (11 May 2011), Accessed 9 August 2014, http://dotearth.blogs.nytimes.com/2011/05/11/confronting-the-anthropocene/?_php=true&_type=blogs&_r=0.

40. Dipesh Chakrabarty, "The Climate of History," *Critical Inquiry* 35 (Winter 2009), 197–222.

41. Paul Crutzen, "The Geology of Mankind," *Nature* 415.6867 (2002), 23.

42. Jason Moore, "Anthropocene or Capitalocene?" Accessed 9 August 2014, http://jasonwmoore.wordpress.com/2013/05/13/anthropocene-or-capitalocene/

43. Dirzo, 401.

44. Mark Nathan Cohen, *The Food Crisis in Prehistory: Overpopulation and the Origins of Agriculture* (New Haven, CT: Yale University Press, 1977).

45. Broswimmer, 12–22.

46. Broswimmer, 24.

47. Wilson, 96.

48. Broswimmer, 9.

49. Clive Ponting, *A Green History of the World: The Environment and the Collapse of Great Civilizations* (NY: Penguin, 1991), 37.

50. William F. Ruddiman, "The anthropogenic greenhouse era began thousands of years ago". *Climatic Change* 61.3 (2003): 261–293.

51. Ponting, 54.

52. Steven Roger Fischer, *History of Writing* (New York: Reaktion Books, 2004), 22.

53. Fischer, 36.

54. Broswimmer, 36.

55. Brian R. Ferguson, "Ten Points on War," Social Analysis, 52.2 (Summer 2008), 32–49.

56. Ponting, 58.

57. Robert Pogue Harrison, *Forests: The Shadow of Civilization* (Chicago, IL: University of Chicago Press, 1992), 17.

58. Neil Roberts, *The Holocene: An Environmental History* (New York: Basil Blackwell, 1992).

59. Joseph Tainter, *The Collapse of Complex Societies* (New York: Cambridge University Press, 1990.

60. J. Donald Hughes, *Environmental Problems of the Greeks and Romans: Ecology in the Ancient Mediterranean* (Baltimore, MD: Johns Hopkins University Press, 2014).

61. Broswimmer, 42.

62. Broswimmer, 42.

63. Broswimmer, 42.

64. J. Donald Hughes, "Ripples in Clio's Pond: Rome's Decline and Fall: Ecological Mistakes?" *Capitalism, Nature, Socialism* 8.2 (June 1997), 117–21.

65. Sally Grainger, ed., *Apicius: A Critical Edition* (New York: Prospect Books, 2006).

66. Tainter, 146.

67. Christopher Columbus, *Select Letters of Christopher Columbus*, R.S. Major, trans and ed. (London: Hakluyt Society, 1870), 5.

68. Columbus, 4.

69. Joel Kovel, *The Enemy of Nature: The End of Capitalism or the End of the World* (New York: Zed, 2007), 38.

70. Quoted in Kovel, 41.

71. Ponting, 179.

72. Broswimmer, 63.

73. Alfred Crosby, *The Columbian Exchange: Biological and Cultural Consequences of 1492* (New York: Praeger, 2003).

74. Richard Grove, *Green Imperialism: Colonial Expansion, Tropical Island Edens, and the Origins of Environmentalism, 1600–1860* (New York: Cambridge University Press, 1996).

75. Grove.

76. Eric Williams, *Capitalism and Slavery* (Charlotte, NC: University of North Carolina Press, 1994).

77. Vandana Shiva, *The Violence of the Green Revolution: Third World Agriculture, Ecology, and Politics* (New York: Zed Books, 1992).

78. John Locke, *Second Treatise on Government*, Chapter 5: Of Property, http://www.constitution.org/jl/2ndtr05.htm

79. Barbara Arneil, *John Locke and America: The Defense of English Colonialism* (New York: Oxford University Press, 1996).

80. Carolyn Merchant, *The Death of Nature: Women, Ecology, and the Scientific Revolution* (New York: HarperOne, 1990).

81. Silvia Federici, *Caliban and the Witch: Women, the Body, and Primitive Accumulation* (New York: Autonomedia, 2004), 65.

82. Vandana Shiva and Ingunn Moser, eds., *Biopolitics: A Feminist and Ecological Reader on Biotechnology* (Atlantic Highlands, NJ: Zed, 1995).

83. David Harvey, *The Enigma of Capital* (New York: Oxford University Press, 2010), 217.

84. Ponting, 186.

85. John F. Richards, *The World Hunt: An Environmental History of the Commodification of Animals* (Berkeley, CA: University of California Press, 2014), 112.

86. Richards, 134.

87. Richards, 131.

88. Broswimmer, 68.

89. Broswimmer, 68.

90. Adam Smith, *The Wealth of Nations*

91. Ponting, 155.

92. For a discussion of the economic and political mechanisms that generate imperialism, see David Harvey, *The New Imperialism* (New York: Oxford University Press, 2003).

93. Roxanne Dunbar-Ortiz, *An Indigenous People's History of the United States* (New York: Beacon, 2014).

94. Broswimmer, 65.

95. Rebecca Solnit, *Savage Dreams: A Journey in the Hidden Wars of the American West* (San Francisco: Sierra Club, 1994).

96. David Zierler, *The Invention of Ecocide: Agent Orange, Vietnam, and the Scientists Who Changed the Way We*

Think About the Environment (Athens, GA: University of Georgia Press, 2001).

97. Barry Sanders, *The Green Zone: The Environmental Costs of Militarism* (Oakland, CA: AK Press, 2009).

98. Naomi Klein, *This Changes Everything: Capitalism vs. The Climate* (New York: Simon and Schuster, 2014).

99. Damian Carrington, "Earth Has Lost Half Its Wildlife in the Past Forty Years, WWF Says," *The Guardian* (30 September 2014), http://www.theguardian.com/ environment/2014/sep/29/earth-lost-50-wildlife-in-40-years-wwf.

100. Adrian Parr, *The Wrath of Capital: Neoliberalism and Climate Change Politics* (Columbia University Press, 2014).

101. It's worth noting that this is also the title adopted by a contemporary coalition of Indigenous and anti-colonial groups from the Global North and South. See Demos, "Climate Control."

102. On conservation's colonial origins and enduring racism, see Grove, *Green Imperialism* and Dowie, *Conservation Refugees: : The Hundred Year Conflict Between Global Conservation and Native Peoples* (Cambridge, MA: MIT Press, 2009).

103. Armiero, *Wasteocene: Stories from the Global Dump* (New York: Cambridge University Press, 2021).

104. Harvey, *The New Imperialism*, 87.

105. United Nations, "Sustainable Development Goals", https:// sdgs.un.org/goals/goal12.

106. IUCN, "IUCN Congress Closes With Call for a Post-Pandemic Recovery Based in Nature" (10 Sept 2021),

https://www.iucn.org/news/secretariat/202109/iucn-congress-closes-call-a-post-pandemic-recovery-based-nature.

107. IUCN, "Motion 101", https://www.iucncongress2020.org/motion/101.

108. Jeff Tollefson, "Global Deal for Nature' Fleshed Out with Specific Conservation Goals" *Nature* (19 Apr 2019), https://www.nature.com/articles/d41586-019-01253-z.

109. "Our Land, Our Nature" Congress, https://www.ourlandournature.org/.

110. The *People's Manifesto for the Future of Conservation* is available at https://www.ourlandournature.org/manifesto.

111. WWF, "WWF's History", https://www.worldwildlife.org/about/history.

112. Quoted in Kashwan et al., "From Racialized Neocolonial Global Conservation to an Inclusive and Regenerative Conservation," *Environmental* 63.4 (2021)," 5.

113. Brenna Bhandar, *Colonial Lives of Property: Law, Land, and Racial Regimes of Ownership* (Durham, NC: Duke University Press, 2018).

114. Gadgil and Guha, *This Fissured Land: An Ecological History of India* (New York: Oxford University Press, 2012), 132.

115. Kashwan, 6.

116. Kashwan, 10.

117. Kashwan, 7.

118. Dowie, "Conservation Refugees".

119. Dowie, Conservation Refugees".

120. Protected Planet, "Executive Report," https://livereport.protectedplanet.net/chapter-1.

121. Schleicher, "Protecting Half of the Planet."

122. E. O. Wilson, *Half-Earth: Our Planet's Fight for Life* (New York: Liveright, 2017).

123. Fiore Longo, "Personal Interview".

124. Simon Counsell, "Anatomy of a 'Nature-Based Solution,'" *REDD Monitor* (16 Apr 2021), https://redd-monitor. org/2021/04/16/anatomy-of-a-nature-based-solution-total-oil-40000-hectares-of-disappearing-african-savannah-emmanuel-macron-norwegian-and-french-aid-to-an-election-rigging-dictator-trees/.

125. Quoted in Counsell, "Anatomy."

126. Counsell, "Anatomy."

127. Kashwan, 7.

128. Guha and Martinez-Alier, *Varieties of Environmentalism*,12.

129. Ibid.

130. On degrowth, see Jason Hickel, *Less is More: How Degrowth Will Save the World* (New York: Penguin, 2021).

131. Kristin Ross, *Communal Luxury: The Political Imaginary of the Paris Commune* (New York: Verso, 2016).

132. WWF, *Living Planet Report 2014*, http://wwf.panda.org/ about_our_earth/all_publications/living_planet_report/.

133. J. B. MacKinnon, *The Once and Future World: Nature As It Was, As It Is, As It Could Be* (New York: Houghton Mifflin Harcourt, 2013).

134. George Monbiot, *Feral: Searching for Enchantment on the Frontiers of Rewilding* (New York: Allen Lane, 2013), 84.

135. Monbiot, *Feral*, 12.

136. Kolbert, *Sixth Extinction*, p. 159.

137. For an overview of the debate around assisted colonization, see John Lascher, "If You Plant Different Trees in the Forest, Is It Still the Same Forest?" *The Guardian* (19 October 2014), http://www.theguardian.com/vital-signs/2014/oct/19/-sp-forests-nature-conservancy-climate-change-adaptation-minnesota-north-woods.

138. Philip Seddon *et al.*, "Reversing Defaunation: Restoring Species in a Changing World," *Science* 345.6195 (2014): 406–412.

139. "Environmentalism and Postcolonialism," in Ania Loomba and Suvir Kaul, eds., *Postcolonial Studies and Beyond* (Durham, NC: Duke University Press, 2005), 233–51.

140. For a deconstruction of such notions of the pristine nature of wilderness, see William Cronon, "The Trouble With Wilderness; or, Getting Back to the Wrong Nature," in William Cronon, ed., *Uncommon Ground: Rethinking the Human Place in Nature* (New York: W. W. Norton & Co., 1995), 69–90.

141. Sergey Zimov, "Pleistocene Park: Return of the Mammoth's Ecosystem," *Science* 308.5723 (6 May 2005), 796–798.

142. Paul S. Martin, *Twilight of the Mammoths: Ice Age Extinctions and the Rewilding of America* (Berkeley, CA: University of California Press, 2005); Josh Donlan, "Lions and Cheetahs and Elephants, Oh My!" *Slate*, August 18, 2005. See also *The Rewilding Institute* at http://rewilding.org/rewildit/ and *Rewilding Europe* at http://www.rewildingeurope.com/.

143. Elizabeth Kolbert, "Recall of the Wild: The Quest to Engineer a World Before Humans," *The New Yorker*

(24 December 2012), http://www.newyorker.com/magazine/2012/12/24/recall-of-the-wild.

144. George Church and Ed Regis, *Regenesis: How Synthetic Biology Will Reinvent Nature and Ourselves* (New York: Basic Books, 2012), 10.

145. On the transformation of life into code, see Eugene Thacker, *The Global Genome: Biotechnology, Politics, and Culture* (Cambridge, MA: MIT Press, 2005).

146. Church and Regis, 10.

147. Nathaniel Rich, "The Mammoth Cometh," *New York Times Magazine*.

148. Church and Regis, 12.

149. Church and Regis, 12.

150. Josh Donlan, "De-extinction in a Crisis Discipline," *Frontiers of Biogeography* 6.1 (2014), 25–28.

151. *Scientific American* Editors, "Why Efforts To Bring Extinct Species Back from the Dead Miss the Point," Scientific American 308.6 (14 May 2013), http://www.scientificamerican.com/article/why-efforts-bring-extinct-species-back-from-dead-miss-point/.

152. Dr. Norman Carlin, Ilan Wurman, and Tamara Zakim, "How To Permit Your Mammoth: Some Legal Implications of 'De-Extinction," *Stanford Environmental Law Journal* 33.3 (January 2014), https://journals.law.stanford.edu/stanford-environmental-law-journal-selj/print/volume-33/number-1/how-permit-your-mammoth-some-legal-implications-de-extinction.

153. Melinda Cooper, *Life as Surplus: Biotechnology and Capitalism in the Neoliberal Era* (Seattle, WA: University of Washington Press, 2008), 27.

154. Cooper, 23.

155. Church and Regis, 4.

156. Melinda Cooper makes this extremely productive link between the monetarist revolution and the growth of research funding for the life sciences. See Cooper, 29–31.

157. David Harvey, *The New Imperialism* (New York: Oxford University Press, 2003), 67.

158. Ilya Prigogine and Isabelle Stengers, *Order Out of Chaos: Man's New Dialogue With Nature* (New York: Bantam, 1984).

159. Cooper, 38.

160. Naomi Klein, *The Shock Doctrine: The Rise of Disaster Capitalism* (New York: Picador, 2008).

161. Stefano Liberti, *Land Grabbing: Journeys in the New Colonialism* (New York: Verso, 2014).

162. Global Justice Ecology Project, *The Green Shock Doctrine* (12 May 2014), http://globaljusticeecology.org/green-shock-doctrine/.

163. Ibid, 8.

164. John Vidal, "How the Kalahari Bushmen and Other Tribespeople Are Being Evicted to Make Way for 'Wilderness,'" *The Guardian* (15 November 2014), http://www.theguardian.com/world/2014/nov/16/kalahari-bushmen-evicted-wilderness.

165. A promising initial step in this direction was taken during the UN Convention on Biological Diversity meeting in 2014, although it met fierce opposition from nations like the US and UK with strong synthetic biology industries.

See "Regulate Synthetic Biology Now: 194 Countries," *SynBio Watch* (20 October 2014), http://www.synbiowatch.org/2014/10/regulate-synthetic-biology-now-194-countries/.

166. George Monbiot, "A Manifesto For Rewilding the World," http://www.monbiot.com/2013/05/27/a-manifesto-for-rewilding-the-world/.

167. Jonathan Watts, "Ecuador Approves Yasuni National Park Oil Drilling in Amazon Forest," *The Guardian* (16 August 2013), http://www.theguardian.com/world/2013/aug/16/ecuador-approves-yasuni-amazon-oil-drilling.

168. Peder Anker, *Imperial Ecology* (Cambridge, MA: Harvard University Press, 2009).

169. Henry Veltmeyer and James Petras, eds., *The New Extractivism: A Post-Neoliberal Development Model or Imperialism of the 21st Century* (New York: Zed Books, 2014).

170. On the case for repayment of climate debt, see Andrew Ross, *Creditocracy and the Case for Debt Refusal* (New York: OR Books, 2014).

171. Wilson, *The Future of Life*, 60.

172. James Sayer, *Why We Can't Afford the Rich* (Chicago, IL: University of Chicago Press, 2015).

173. See the Global Commons Institute website: http://www.gci.org.uk/.

174. Sara Jerving, Katie Jennings, Masako Melissa Hirsch and Susanne Rust, "What Exxon Knew About the Earth's Melting Artic," *Los Angeles Times* (9 October 2015), http://graphics.latimes.com/exxon-arctic/.

175. Camilo Mora et. al., "The Projected Timing of Climate Departure from Recent Variability," *Nature* (9 October 2013).

176. Peter H. Gleick, "Water, Drought, Climate Change, and Conflict in Syria," *Weather, Climate, and Society* 6.3 (July 2014), 331–340.

177. Parenti, *Tropic of Chaos*.

178. Erik A. Beever, Chris Ray, Jenifer L. Wilkening, Peter F. Brussard, Philip W. Mote, "Contemporary climate change alters the pace and drivers of extinction" *Global Change Biology*, 2011; DOI: 10.1111/j.1365-2486.2010.02389.x

179. Broswimmer, 7.

180. Cormac Cullinan, *Wild Law: A Manifesto For Earth Justice* (New York: Chelsea Green, 2011).

181. Kolbert, 266.

182. Kolbert, 266.

183. Fredric Jameson, "Future City," *New Left Review* 21 (May– June 2003), http://newleftreview.org/II/21/fredric-jameson-future-city.

184. Klein, *This Changes Everything*.

185. Broswimmer, 7.

ABOUT THE AUTHOR

Born in South Africa during the apartheid era, Ashley Dawson is currently Professor of Postcolonial Studies at the Graduate Center, City University of New York and the College of Staten Island. His previous books include *Extreme Cities: The Peril and Promise of Urban Life in the Age of Climate Change* and *People's Power: Reclaiming the Energy Commons.* A member of the Social Text Collective and the founder of the CUNY Climate Action Lab, he is a longtime climate justice activist.

Love in the Anthropocene

Dale Jamieson and Bonnie Nadzam

ISBN 978-1-939293-90-9 paperback
ISBN 978-1-939293-91-6 e-book
218 pages

"Dale Jamieson and Bonnie Nadzam cause us to think—and to feel—
what life will be like in a future where nothing is left that is spontaneous,
accidental, or uncontrolled. A beautiful—and frightening—book."

—Naomi Oreskes,
professor, history of science, Harvard;
author, *Merchants of Doubt*

OR Books
www.orbooks.com

Welcome to the Greenhouse

Gordon Van Gelder, Editor

ISBN 978-1-935928-27-0 paperback
ISBN 978-1-935928-26-3 e-book
336 pages

"As the planet heats up, almost certainly some regions will experience more intense droughts, but which regions, exactly, and how intense will those droughts be? ....The greatest unknown of all is, of course, how people, collectively, will respond. Will they be chastened? Genocidal? Or will they just muddle along, behind growing seawalls and shrinking coasts? Science—even social science—can't answer questions like this, which is why we turn to science fiction. Welcome to the Greenhouse!"

—Elizabeth Kolbert

OR Books
www.orbooks.com